CW00664543

'A ruthless demolition of the illusions that cover up years of underinvestment in Britain's defence. Disturbing, compulsory reading for anyone who cares about European security in the twenty-first century.' — Jeremy Bowen, International Editor, BBC News

'An extremely informed and thought-provoking critical examination of Britain's current strategic posture, and a clarion call to action. Offering a roadmap for the UK's immediate and long-term strategic direction, it is full of wisdom, insight and clear thinking. A must-read for all Britain's leaders—political, diplomatic and military.' — James Holland, author of *The Savage Storm* and *Normandy '44*

'In liberal democracies, long-term decision-making requires popular support. Our history and geography gave us responsibilities as well as powers. A nuclear power and a key player in NATO, Britain has for too long confused values and interests, and failed to align them.' — Baroness Gisela Stuart, Chair of Wilton Park

'An exceedingly compelling, forthright and alarming assessment of the growing disconnect between the ambitions of Global Britain and the limits of its reduced military capacity at a time of mounting security threats to the UK and to its interests in Europe and around the world. A hugely important book by two highly respected practitioners and thinkers.' — General (Ret.) David Petraeus, former Commander of the Surge in Iraq, US Central Command, and Coalition Forces in Afghanistan; former director of the CIA; and co-author of the *New York Times* best-selling book *Conflict: The Evolution of Warfare from 1945 to Ukraine*

'A full-throated warning about what happens if Great Britain fails to develop the necessary strategy and strengthen the essential alliances required for long-term security and prosperity. But what makes this book special and important are its achievable recommendations for designing such a strategy.' — Lieutenant General (Ret.) Ben Hodges, former commanding general of US Army Europe

'A passionate call from two prestigious authors for the UK to regain its traditional role of strategic and military leadership in Europe and the trans-Atlantic theatre, in the face of an impending Russian threat to our democracies and values. A must-read for all.' — Giampaolo Di Paola, former Italian minister for defence and former chairman of the NATO Military Committee

'There are many who can see strategic decay and describe it in exquisite detail, but precious few who can provide the kinds of insights and prescriptions essential for the UK to ready itself for the kind of war we'll actually face. Thankfully, there are few authors better qualified to educate us on the realistic strategy demanded by our current circumstances and thrust upon us by today's evolving technologies. *The Retreat from Strategy* should be required reading at the war colleges of the NATO nations, and more broadly of the community of democracies.' — General (Ret.) John R. Allen, United States Marine Corps

THE RETREAT FROM STRATEGY

DAVID RICHARDS
JULIAN LINDLEY-FRENCH

The Retreat from Strategy

Britain's Dangerous Confusion of Interests with Values

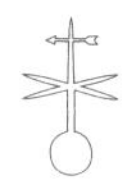

HURST & COMPANY, LONDON

First published in the United Kingdom in 2024 by
C. Hurst & Co. (Publishers) Ltd.,
New Wing, Somerset House, Strand, London, WC2R 1LA

Distributed in the United States, Canada and Latin America by
Oxford University Press, 198 Madison Avenue, New York, NY 10016,
United States of America.

A Cataloguing-in-Publication data record for this book
is available from the British Library.

ISBN: 9781911723677

This book is printed using paper from registered sustainable
and managed sources.

www.hurstpublishers.com

Printed and bound in Great Britain by Bell & Bain Ltd, Glasgow

Though much is taken, much abides; and though
We are not now that strength which in old days
Moved earth and heaven; that which we are; we are,
One equal temper of heroic hearts,
Made weak by time and fate, but strong in will
To strive, to seek, to find and not to yield.

"Ulysses", Alfred, Lord Tennyson

For the men and women of the British Armed Forces

CONTENTS

ABOUT THE AUTHORS

General Lord David Richards

A member of the Alphen Group, General David Richards led operations in East Timor, Sierra Leone, and Afghanistan. He is probably best known for his command in Sierra Leone in 2000 when he interpreted his orders creatively to achieve more than was at first thought possible, ensuring the ultimate defeat of the Revolutionary United Front (RUF) rebels and the avoidance of much bloodshed in the capital Freetown. He went on to command the NATO campaign in Afghanistan during the Alliance's expansion of responsibility across the whole country. Having first commanded the British Army, in 2010 he became Chief of the Defence Staff, the professional head of Britain's armed forces and their strategic commander as well as the Prime Minister's military adviser and a member of the National Security Council. He retired in July 2013.

His UK operational awards include a Mention in Despatches, Commander of the British Empire, Distinguished Service Order and Knight Commander of the Bath. He is the first officer to receive an operational knighthood since World War 2. In 2011 he received the annual Churchillian Award for leadership. He was created Baron Richards of Herstmonceux in February 2014 and now sits in the House of Lords. Amongst other appointments he is an Honorary Fellow of both King's College London and Cardiff University and until recently he was Executive Chairman of Equilibrium Global, the geo-strategic advisory company, and a Visiting Professor of Exeter University. He is actively involved with several charities, especially the Royal Commonwealth Ex-Services League of which he is the Grand President. His autobiography *Taking Command* was published in October 2014. He was President of the high-level trilogy of Future War conferences which took place in October 2023.

Professor Julian Lindley-French MA (Oxon.), MA (Dist.), PhD

Julian Lindley-French is Chairman of the Alphen Group, Senior Fellow at the Institute for Statecraft in London, Director of Europa Analytica in the Netherlands, and a Fellow of the Canadian Global Affairs Institute. He was educated at the University of Oxford, UEA and received his doctorate from the European University Institute in Florence. He has many published books to his name, as well as major reports and articles and has held three professorial chairs, including Eisenhower Professor of Defence Strategy at the Netherlands Defence Academy. He was also Strategic Advisor to two Chiefs of the British Defence Staff and has held senior policy, operational and project positions for the UK, NATO, EU and UN in London, Brussels, Geneva, New York and elsewhere. He was also formerly Director of both the International Training Course and the European Training Course at the Geneva Centre for Security Policy. In 2022, he published *Future War and the Defence of Europe* (Oxford: Oxford University Press & Stuttgart: Kosmos) with US General (Ret.) John R. Allen and US Lieutenant General (Ret.) Ben Hodges. In 2022, he drafted the Shadow NATO Strategic Concept, and in March 2023 presented "A Comprehensive Strategy for a Secure Ukraine" to the European

Parliament. In May 2023, he published a third edition of his successful book *North Atlantic Treaty Organization* (New York: Routledge). He is also Director of the high-level Future War trilogy of conferences.

ACKNOWLEDGEMENTS

We are both immensely grateful to the senior people who gave of their time and effort to support this book. These included a former Prime Minister, Sir Tony Blair, and two former Secretaries of State for Defence, Lord George Robertson and Lord Des Browne. We must also thank several former NATO Assistant Secretaries-General and a Deputy Assistant Secretary-General, most notably Robert Bell and Gordon Davis—close allies and good friends. Professor Rob de Wijk, founder of the Hague Centre for Security Studies was, as ever, most helpful. Vice-Admiral (Ret.) Peter Hudson and Air Marshal Ed Stringer played a vital role in the chapter on the British future force, whilst Vice Admiral Sir Anthony Dymock afforded us an equally vital reality check. James Joye Townsend Jr kindly agreed to write the foreword. For many years, Jim was the Pentagon's Deputy Assistant Secretary of Defense for Eurasia and has had a ringside American seat at Britain's retreat from strategy.

We would also like to thank fellow members of the Alphen Group for their support in considering the optimal British future force. General (Ret.) John Allen, Dr Hans Binnendijk, Professor Yves Boyer, Brig. Gen. (Ret.) Robbie Boyd, Major-General (Ret.) Gordon Davis, General (Ret.) Sir James Everard, Air Marshal (Ret.) Sir Christopher Harper, Lt. General (Ret.) Ben Hodges,

Dr Rich Hooker, Diego Ruiz Palmer, Brigadier General (Ret.) Mick Ryan, General (ret.) Sir Rupert Smith and Professor Rob de Wijk.

Last, but by no means least, we pay tribute to our respective wives Caroline Richards and Corine Schouten who have had to put up with us banging on about Britain and strategy for over a year—and not for the first time.

FOREWORD

BY JAMES JOYE TOWNSEND JR, FORMER US DEPUTY ASSISTANT SECRETARY FOR DEFENCE

Quantity has a quality all its own.

Attributed to Josef Stalin

Devising a strategy to impose order over chaos is a difficult undertaking given how few strategies succeed unscathed, and sometimes come at a political cost. "Muddling through" the chaos is far easier and safer than designing an exquisite strategy just to watch it melt away like a sandcastle on a storm-tossed beach. However, muddling through is not an option today. The West and the nations that make it up need a strategy to quickly rebuild militaries into modern fighting forces that can deter an aggressive Russia that once again stalks Europe. While NATO has the plans to defend the Alliance, allies must fill them out with forces. And soon.

The thirty-year vacation from European defence enjoyed by the West since the end of the Cold War is over, and has been for some time. Like the efforts by British governments in the 1930s to avoid anything that might bring about another world war, seventy years later leaders who lived through the Cold War swore to do anything to avoid its return. While such an aspiration is understandable, the West was slow to recognize when it was failing, beginning with the Russian invasion of Georgia in 2008. Afterwards, the West's doubling down by resetting relations with Russia only showed Putin how eager they were to avoid conflict with him.

After the Georgia invasion, it was rational to make sure no opportunities were lost to build a Europe "whole, free and at peace"—but rational only if it was accompanied by hedging, by the West buying insurance against failure by simultaneously increasing defence spending and strengthening deterrence. However, there was no hedging, especially after wars in Afghanistan and Iraq and the financial crisis of 2008 led to the creation of lighter forces specialising in counter-terrorism, smaller force structures and movement away from readiness to fight high intensity conflicts. Instead, the vacation from European defence continued based on the faulty assumption that Moscow shared the West's desire not to return to the Cold War. The rise of an aggressive China became a higher priority for the US and led to its "pivot to Asia," funded in part by the Pentagon's decision to "take risk" in Europe by withdrawing forces from there to help pay for the build-up in the Indo-Pacific.

It took the brutality of Putin's second invasion of Ukraine in 2022 to truly convince the West that the era of building a new Europe was over and aggression once again walked astride the continent. While NATO was quicker to return to the defence of Europe, the nations were caught flatfooted, including the United States. NATO can develop the best plans for the defence of Europe, but without strong military contributions from the allies to fill out that plan with boots on the ground, the plans are hollow. The scramble is now on in most Allied nations to make up for thirty years of neglect by returning strength to atrophied military muscle. This scramble becomes urgent if Ukraine falters and is overtaken by a large and recapitalized Russian Army that Putin may use elsewhere.

As in 1938, when Europe found itself facing down another aggressive dictator, the UK must be a leader in this effort to rebuild European deterrence, along with France and Germany. Should the United States become overstretched as it meets simultaneous threats in the Indo-Pacific, as well as Iranian sponsored turmoil in the Middle East, this troika will have to spearhead not just a military response to deter Russian aggression, but leadership within the Alliance to ensure all the allies man the ramparts. This time, the hedging must be against the potential that the US cannot be present in Europe with its full military capacity.

From a Washington perspective, what do we see is needed from the UK? More... more of everything. It is not a choice between conventional or nuclear forces, it must be both. It is a UK military with the full range of capabilities, from nuclear forces to special forces, from more aircraft to more armour to more ships, to forces at a higher state of readiness, including forces already deployed to the frontier with Russia. These forces must also reside on the leading edge of military technology, such as we see being deployed in Ukraine, and have logistics that include large stocks of ammunition. These are not new requirements: the fighting in Ukraine shows us where we are deficient, and there are many areas that need attention, not just in the UK but in the US and across the Alliance. These deficiencies are well known in London too and their resolution rests at the feet of politicians; the West waits to see if they will rise to the occasion. The US always wants the British military alongside it in a fight, we just need more of them.

Today, when the velocity and simultaneity of crises stress every sinew of our national security structure, muddling through is not an option. Given the challenges ahead, described in this book by two leading experts in the field of military strategy, there is no time to waste with muddled thinking. Strategies should already be underway in capitals to rebuild militaries that have been neglected for too long, to refill ammunition bunkers, replenish the arsenals and train a new generation of soldiers to once again man the Western ramparts against a Russian aggressor. As this is being written, a great NATO military plan is being completed that provides those ramparts across Europe. But to man them, NATO depends on the nations to bring forward the military capability to do so, and that's where each nation's strategy will need to do the job.

The situation that the UK finds itself in has been eloquently described by an MP in the House of Commons:

> Now the victors are the vanquished, and those who threw down their arms in the field and sued for an armistice are striding on to world mastery. That is the position, that is the terrible transformation that has taken place bit by bit. I rejoice to hear from the Prime Minister that a further supreme effort is to be made to place us in a position of security: Now is the time at last to rouse the nation. Perhaps it is the last time it can be roused with a chance of preventing war, or with a chance of coming through to victory should our efforts to prevent war fail. We should lay aside every hindrance to endeavour by uniting the whole force and spirit of our people to raise again a great British nation standing up before all the world, for such a nation, rising in its ancient vigour, can even at this hour save civilisation.

These words were spoken on 24 March 1938, and the speaker was Winston Churchill. We have no time to lose.

James Joye Townsend

PREFACE

THE SLEEPWALKERS

I am a big believer in the importance of strategy. This is true on many levels but particularly at a state level. In the complex geopolitical dynamics emerging in the world today, with a small number of very big powers and a larger number of medium-level powers, most states risk being lost at a global level without the right strategy. This is particularly true for the UK. Strategy requires analysis of the policy which is the object of the strategy and then a clear idea of how to achieve it: what problem are you trying to solve and what are the elements of a solution.

Sir Tony Blair, in response to the book's questionnaire

The Pledge

The July 2024 general elections heralded a new British government. Such moments are always useful to consider Britain's place in the world. This book does precisely that by focussing on the utility and relevance of British military power as one instrument amongst several available to London in the eternal struggle for influence, liberty, and freedom. The book's overarching aim is to help government far more effectively balance the ends, ways and means of defence. There is good news and bad news. The good news is that on 23 April 2024, St George's Day, Prime Minister

Rishi Sunak pledged that Britain would increase its defence budget from just over 2% of GDP to 2.5%. By 2030, according to Sunak, £23 billion would be added to the military budget, whilst he claims that overall "defence" would receive an additional £75 billion, bringing the entire defence budget up to £87 billion a year. This would make Britain comfortably the biggest defence investor in Europe.

Such investment would at least help cover the existing hole in the defence investment plan and replace badly depleted stocks of ammunition and vital logistics, so long as such capacity is not shipped to Ukraine. Although supporting Ukraine in the now long and costly Russo-Ukraine War is part of British defence strategy, Britain has not articulated clear war aims there. In March 2024, the House of Commons Public Accounts Committee revealed a £16.9 billion deficit in the Defence Equipment Plan, the largest such mismatch between means and ways since the Ministry of Defence first published its plans in 2012. The deficit is likely to be closer to £20 billion. The Committee did not pull its punches: "Either the Ministry of Defence must be fully funded to engage in operations whilst also developing warfighting readiness; or the government must reduce the operational burden on the Armed Forces."[1] In real terms Sunak's advertised hike would have at best only matched the level of defence expenditure in 2010 at the end of the last Labour government. And, whilst there are many lessons from the Russo-Ukraine War that are of limited relevance to Britain's armed forces, the need for a baseline mass of forces is one because in any war "hard yards" will always have to be won and there is only so much that technology can do, at least for the foreseeable future.

A further £10 billion was to be invested in re-building Britain's own domestic defence industry. This evokes memories of the pre-war Shadow Factory Plan and is to be welcomed given it would afford industry more secure, longer-term contracts at some scale and would improve what have been lamentable fielding times of equipment. The Australia, UK, US (AUKUS) submarine programme (SSN-AUKUS) is already doing this, albeit at great cost. A new Defence Innovation Agency is also planned; modelled on the US Defense Advanced Research Projects Agency (DARPA), which would look at how best to exploit emerging and disruptive technologies (EDT), not least autonomous drones, both airborne and sub-sea, and a possible future ground-based air defence system for Britain not dissimilar to Israel's "Iron Dome." This agency would also support Space Command and the British, Italian, Japanese consortium as it develops a 6th Generation (6G) fighter. More cutting-edge 5G equipment was called for, such as more F-35 strike aircraft for the Royal Air Force (RAF), whilst twenty-two new ships and submarines are being constructed for the Royal Navy (RN). Above all, more would be invested in recruitment and retention, which will be vital. Even elite formations which form the speartip of Britain's expeditionary forces, the Parachute Regiment and the Royal Marines, have been hollowed out of late. The Paras have been reduced from 2,200 troops (still tiny) in 2016 to 2,030 in October 2023. The Royal Marines were similarly reduced from 7,110 in 2016 to 6,040.

The bad news is that such investment is probably too little, too late, and not as much as then Prime Minister Sunak claimed. This has been the essential story of British defence policy and strategy since the end of the Cold War, and will not stop the ends, ways and means crisis faced by the British armed forces, whoever is in Downing Street. Former Armed Forces Minister James Heappey said in April 2024 that the additional funding would do nothing to increase the size of the British armed forces, which is meant to be one of two strategic reserves available to NATO's Supreme Allied Commander, Europe (SACEUR). Worse, for thirty years London has ignored deterrence, defence and the warfighting forces such missions need to be credible in preserving the peace. Counterinsurgency operations or COIN have been all the rage. The Americans solve this tension at scale by spending 3.3% (£670 billion) per annum on defence and buying redundancy of capability. Still, much of the proposed increase was also driven by the need for London and other allies to ease growing pressure on the American armed forces, and they need to because whilst Global Britain is a political fantasy, Global America is not.

On 24 April 2024, during an interview with David Richards, the BBC's "PM" programme presenter Evan Davis pointed out that the actual increase in defence spending will be closer to £20 billion than £75 billion and just over £6 billion per annum beyond 2027. As Richards said, "the funding will help achieve plans already set in motion," but is unlikely to be enough given some big spending decisions must be made that are still not fully funded.

In December 2023, the then Secretary of State for Defence Grant Shapps said,

> we will spend 2.5 per cent as soon we can. I've spoken previously about the need to go to 3 per cent and I don't resile from that at all. I think we need to go to 3 per cent and possibly higher... and I think the rest of the West is going to have to do the same.[2]

3% GDP per annum on defence by 2030 would certainly seem a more reasonable goal, not least given the scope and nature of the threats behind Sunak's April 2024 statement. Even if all the non-American NATO allies spent just 2.5% GDP on defence by 2030 that would mean an additional £140 billion for the Alliance, which would go a long way to realising a future NATO in which the Europeans provide 50% of NATO capabilities. This is probably the least the Americans will demand.

As always, the strategic devil is in the political detail. Sunak proposed spending an additional £4.5 billion on defence in financial year 2028–29 and to do so by cutting the civil service, whilst the rest was to have been funded from research and development budgets. This would be an achievement given the civil service has grown exponentially since 2010. It is also an old political trick to hypothecate savings yet to be made as the basis for additional funding elsewhere in government, and it rarely works. It was also easy for Sunak to make such claims when he knew he was unlikely to be in power come 2025, let alone 2030. He was right, indeed, much of the Sunak statement smacked of politics, designed to put the Labour Party in a difficult position on the eve of a general election. Sacrificing long-term strategy for short-term politics has a long and ignoble tradition in London.

There is one further observation that the Sunak statement warrants. In 2021, the government launched the Integrated Review of Security, Defence, Foreign Policy, and Development. By 2023, that had been superseded by the Integrated Review Refresh. Both documents were impressive works of analysis but little more because their implied strategy was always prey to the politics of the moment. Now there will need to be a refresh by the Labour government. This begs a further question: are such reviews worth the paper they are written on?

2% GDP, 2.5% GDP or whatever per cent GDP on defence is not the issue, not least because so much defence investment is wasted. Britain needs a national strategy for defence that matches the political ends vital to secure Britain's critical interests, the means defence needs to play its vital role in meeting those ends, and the ways to apply the means as efficiently and effectively as possible. For, as Richards said in the "PM" interview, London still needs to answer the one vital question it has for decades dodged and which this book seeks to answer: "What does Britain want to achieve in the world?"

The Questions

The Oxford English Dictionary defines "strategic" as "dictated by, and serving the ends of strategy."[3] So, what is the strategy, why and to what end? On 15 January 2024, Grant Shapps gave a speech entitled "Defending Britain from a more dangerous world," which encapsulated everything that is wrong about contemporary British grand strategy and defence policy. Shapps said,

> Today, Russia and China have been joined by new nuclear, and soon to be nuclear, powers. North Korea is promising to expand its own nuclear arsenal. And then there is Iran, whose enriched uranium is up to 83.7%, a level at which there is no civilian application. Back in the days of the Cold War there remained a sense that we were dealing with rational actors. But these new powers are far more unstable, and irrational. Can we really assume the strategy of Mutually Assured Destruction that stopped wars in the past will stop them in future, when applied to the Iranian Revolutionary Guard or North Korea? I am afraid we cannot.

Shapps continued,

> [T]he world is becoming more dangerous and has done in recent years. But the other threats that plagued the start of the 21st century haven't gone away. The spectre of terrorism and threats from non-state actors, as October 7 showed, still haunts the civilised world. Put it all together, and these combined threats risk tearing apart the rules-based international order—established to keep the peace after the Second World War. Today's world then, is sadly far more dangerous.

All well and good, but Shapps then went on to say,

> Over the last decade this government has made great strides to turn the Defence tanker around. The refreshes of the Integrated Review and Defence Command Paper have been instrumental in ensuring Britain is defended in this more dangerous world. We've uplifted our defence spending—investing billions into modernising our Armed

> Forces and bringing in a raft of next generation capabilities, from new aircraft carriers to F35s; from new drones to Dreadnought submarines; from better trained troops; to the creation of a national cyber force. And when the world needed us, we have risen to the moment ... Today, for the very first time this government is spending more than £50bn a year on Defence in cash terms, more than ever before. And we have made the critical decision to set out our aspiration to reach 2.5% of GDP spent on defence. And as we stabilise and grow the economy, we will continue to strive to reach it as soon as possible. But now is the time for all allied and democratic nations across the world to do the same. And ensure their defence spending is growing. Because, as discussed, the era of the peace dividend is over.[4]

Not surprisingly, His Majesty's Treasury pushed back. Britain may spend more in cash terms than at any time in the past, but the pound is worth less than at any time in the past due to inflation, or more precisely defence cost inflation which runs at least 3–4% higher than price inflation, and has reached as high as 8%. As one very senior and very recently-retired former official at the Ministry of Defence said to one of the authors of this book, "The British defence budget is simply too small for the force—both conventional and nuclear—it seeks to generate."[5]

Shapps reinforced the pretence by highlighting the new Dragonfire laser air defence system as proof of British future force fighting power. It is an impressive breakthrough in laser weaponry that the Dragonfire Consortium have pioneered, but when will it be deployed, in what numbers, and what are its vulnerabilities? However, perhaps the most telling phrase concerned reliance on allies. Shapps said,

> In a complex world, no nation can afford to go it alone, so we must continue strengthening our alliances so the world knows they cannot be broken. Defence is in many ways the cornerstone of our relations across the world. Our world leading Armed Forces, cutting-edge

> industrial base, and willingness to support our allies is the reason why Britain is the partner of choice for so many. And among our partnerships, NATO remains pre-eminent. 75 years after its foundation, today NATO is bigger than ever. But the challenges are bigger too. That's why the UK has committed nearly the totality of our air, land, and maritime assets to NATO.

This is hyperbole pure and simple.

One baleful deduction that we will examine is that unless government can find additional money—significantly more than 2.5% GDP—Britain can either be a powerful conventional military actor or a credible nuclear power, but not both. Again, Britain's armed forces face an ends, ways and means crisis with no clear vision or narrative to underpin scale or purpose. This is why the book focuses on the lost relationship between British grand strategy and defence strategy and how the former should drive the scale of ambition, size, and scope of the latter. The true test of any command structure, be it government, military, civil, or commercial, is how it stands up to the intense pressure of a crisis. The official British enquiry into the COVID-19 pandemic has revealed that far from standing up to pressure, the British system of government came close to collapsing. Established policy and considered strategy and structure for dealing with such an emergency was jettisoned in favour of ad hoc decisions by a small elite that often disagreed about how to best apply a broken system to best effect. As former Secretary of State for Health, the Rt. Hon. Matt Hancock MP said on 30 November 2023, the government at the time suffered from a "toxic culture."[6] If government effectively collapsed during a pandemic, it is hard to believe it could withstand the pressures of war. With hollowed-out government, hollowed-out capabilities, and little latent redundancy to cope with shock, Britain today is replete with vulnerabilities and frailties.

The Book

It is with this political background as context that this book offers three critical messages. First, there is a vital need to balance the ends, ways and means of grand, national and defence strategy. If the ends, ways and means of British strategy remain as out of sync as they are now disaster beckons. It is better for a modest London to aspire to a modest but affordable grand strategy, in which Britain's strengths and weaknesses are properly understood and acknowledged, than "winging" it in the hope that everything will be all right on the night. It will not. For Britain that means a NATO-focused Euro Atlantic defence strategy rather than aspirations for a "Global Britain" that takes on the Chinese in the Indo-Pacific. Those days are over.

Second, there is a critical need to create a "system of systems" approach across government in London capable of devising and executing British strategy. The current system of government, especially the National Security Council (NSC) and the Ministry of Defence (MoD), is totally inadequate for undertaking such a critical role, especially as a strategic headquarters for which the MoD will be central to British strategy going forward.

Third, London must create a grounded, proven, and robust mechanism for managing crises much better in real time than is the case today. The COBRA (Cabinet Office Briefing Rooms) system is hopelessly outdated and utterly incapable of devising and overseeing the execution of coherent strategies. The National Security Council must be able to run all aspects of Britain's response during a rapidly developing crisis, from strategic command, through intelligence, delivery and evaluation, and logistics.

The Retreat from Strategy is thus about how Britain found itself in such a mess, and how to get out of it. It is about the retreat from strategy, grand strategy and the national and defence strategies that flow from it, and how to restore a system for the use of state power that was once renowned for its craft and resilience. There is little understanding at the apex of power in London about the utility of soft and hard power, and their considered application in an increasingly unsafe world. Strategy has become politics by another name and thus little more than strategic pretence—the short-term dressed up as the long-term, the irrelevant offered as the substantive, and the management of irreversible decline. Not only is Britain's "managed decline" being very poorly managed, but Britain is not in fact declining. Rather, it is recovering its more normal place after some three centuries of exceptionalism. Some states cope better than others with such a change in status. Russia has not and is not coping well. The essential problem for Britain is not post-imperial hubris but the loss of faith in Britain as a power and the distinctive "strategic brand" that goes with it. More precisely, London's political and bureaucratic elite have lost faith in Britain, its people and themselves, condemning a major power to the geopolitical wilderness and making Europe and the wider world more dangerous as a result. As Sir Tony Blair put it,

> In the context of a risen China, a powerful US and a host of other medium-size powers, the UK risks significantly reduced global relevance without the right plan. The right strategy will give the UK the capability to project and protect its values overseas. Where the question of "managed decline" comes in, in my view, is in assessing what a Britain without such a plan looks like. Without a clear focus on our strengths, how to build on them, what partners we need and so on, we risk an inevitable decline in our global role.[7]

Britain is thus sleepwalking towards disaster. There are many causes of this failure of leadership and imagination: repeated crises since 2008 that have shaken an already fragile Britain to its social and political core; political incompetence; strategic pretence; an increasingly ideological London bureaucracy that wants to see Britain, Europe and the world as it would like them to be, rather than as they are; a Treasury that recognises only as much threat as it believes Britain can afford; and an over-reliance on a changing America that increasingly sees Britain as just yet another burdensome European ally. The halcyon days, which were not as halcyon as many Britons like to pretend, of Churchill's "special relationship" with Roosevelt's America are now ancient history in geopolitical terms. The Americans always have and always will act in pursuit of their own interests.

The crises that have hit Britain in this century of crises have also weakened the West and all the liberal democracies. As Professor Rob de Wijk writes,

> Western Europe has become post-modern because of long-term prosperity, security, and the absence of existential challenges. This allowed governments to focus on the further development of the welfare state. Security was no longer seen in terms of territorial integrity, but in terms of humanitarian security. The armed forces therefore became a force for good. No global challenge requires a long-term vision and investment in, for example, the armed forces. This is characteristic of the post-cold war era. But that era is now coming to an end. Moreover, in the post-Cold War era, the West was unquestionably the strength. This meant that politicians did not have to worry about the rest of the world. Strategic thinking faded into the background. As a result of the geopolitical shifts, we are now on the eve of a new paradigm shift that makes it necessary for strategic thinking, and in fact a grand strategy shared by the entire West (G7).[8]

9/11, Afghanistan, Iraq, the 2008–2010 financial and banking crash, Brexit and COVID have all contributed in their own unique ways to Britain's precipitous relative decline. What links them all is a lack of leadership at the political level reinforced by partial and often incompetent policy and strategy poorly implemented by the High Establishment, that uniquely British combination of Whitehall and Westminster insiders who assume the privilege of power.

George Robertson, the former Secretary of State for Defence and NATO Secretary-General frames the challenge with four V's:

> There is a Volatility in geopolitics unprecedented in modern history. Political churn, events like Brexit, movements like populism, organisations like al-Qaeda and ISIS, phenomena like migration flows, people like Donald Trump and Boris Johnson (and Jeremy Corbyn too). Combine all that with a Velocity of change unlike anything in history. As Canadian Prime Minister Justin Trudeau put it, "The pace of change has never been this fast, and yet it will never be this slow again" ... That Volatility of events combined with the Velocity of change has produced new Vulnerabilities we now see playing out in our daily lives ... My fourth V is a Vacuum of leadership at a time when we desperately need some strategic thinking in this new, complex world.[9]

Some blame the current state of British foreign, security and defence policy on the COVID crisis, and there's no question the pandemic had a savage impact on a fragile British economy. However, Britain's malaise not only goes far deeper than any one crisis, but it also goes to the heart of government in London. The COVID crisis simply revealed the appallingly dysfunctional state of government machinery, its inability to deal with big threats and challenges, and the profound confusion of values with interests. Jerry Pournelle, an American science-fiction writer, once crafted what he called "The Iron Law of Bureaucracy." In any bureaucratic organisation there will be two kinds of people: those who work to further the actual goals of the organisation, and those who work for the organisation itself. The Iron Law is that organisations are inherently and inevitably vulnerable to capture by elite activists and it is they who once in control write the rules and determine the goals of that organisation irrespective of any political or other forms of accountability.[10] In Whitehall, the British interest is something of a tainted concept, with demands that Britain follow an "ethical" foreign and security policy even at its own expense—the most obvious example being the commitment to a Net Zero Strategy by 2035 that is neither politically nor technologically feasible.

Nor will Britain again find its place in the world by wallowing in nostalgia. This book firmly embraces much of the "New Thinking" about the need for London to reflect the society it serves if Britain is to become what it should become, a country where all the talents can flourish and contribute to the national story irrespective of race, gender, and sexual orientation. It also argues that intolerant woke thinking that seems to have taken hold across much of government is anything but New Thinking and is in danger of neutering Britain's right to engage in pursuit of its own legitimate security and defence interests. Rather, it is as exclusive and discriminatory as the class-based order of old and thus prevents the return to sound strategy and strategic thinking. Ideologues see the role of government as being to serve their ideology rather than the people. Those that believe the very idea of Britain as a "power" is distasteful and are burdened by historical guilt are simply strategically illiterate. Since such ideologues are not dispassionate in their view of twenty-first century Britain, they are as prone to groupthink as the conservative bastions of the past.

Britain has been, continues to be, and if it is well-led will continue to be a force for good in the world. However, for any democratic state to compete and flourish in a competitive world those in power must first and foremost understand the distinction between values and interests. That does not mean historical myth can pass unchallenged, far from it. Julian Lindley-French is an Oxford historian and the first lesson he learnt at Oxford University is that one cannot and must not judge the past through contemporary mores. Capturing history to serve contemporary ideology is the abandonment of history itself, the very antithesis of sound historiography, something which too many proselytizing British academics seem to have forgotten.

Why does it matter if a major power abandons interests for values? In the absence of a Leviathan to guarantee the peace, states assess and judge each other's interests as the basis for policy. If a powerful state abandons its interests in favour of values, however noble they may be, the already anarchic international system of states which is always prone to instability is liable to become more so. Institutions such as the United Nations and the European Union in their purest form are simply mechanisms for preventing extreme state action through the ordered and orderly conduct of international affairs. If a state is driven by values or ideology it is very hard for others to assess, and as history attests such states can quickly become unstable, aggressive or both. States need to understand Britain's vital and essential interests so that diplomacy can function, and international affairs remain as predictable as possible. This is particularly important for allies: constant virtue-signalling, in which the gap between what London says it can do and reality, is rendering Britain an increasingly unreliable partner.

In 1933, responding to Colonel J.F.C. Fullers' pamphlet on Combined Arms, Erwin Rommel wrote, "The British write some of the best doctrine in the world, it is fortunate their officers do not read it." Pretty much the same can be said for British strategy documents and Britain's political leaders. One of the many paradoxes of contemporary Britain is that, given its history and experience, if there is one country that could create an orderly, predictable international system it is Britain, but only if it makes hard choices, something British governments of all persuasions find very difficult to do. Unfortunately, a political culture has been established at the highest levels of government in London that eschews real choice for the appearance of choice. Thus whilst "strategies" abound in London, the strategic is all too often absent. In 2023, the British government committed yet another strategic sleight of hand when it announced a marked increase in the defence budget whilst at the same time slashing further an already hollowed-out and inadequate British Army, even during a major war in Europe. That decision was the impetus for this book, which aims to enable an intelligent and informed readership to better understand just how Britain finds itself in a position in which the link between policy and effect—strategy—is broken and the instruments of power so weak.

Finally, the fundamental question the book seeks to answer is this: what role should *this* Britain aspire to play now and into the future to preserve its liberty given Britain's still very considerable wealth and power, its strategic and political interests, and its still-capable armed forces? Not the past Britain of empire but this Britain—a top-ten world economy and nuclear power with seventy million reasonably-educated and well-connected people, with a historical tradition of power and influence in a relatively efficient geographical space off the northwest coast of a still-affluent continent.

SCENARIO

BRITAIN DEFEATED

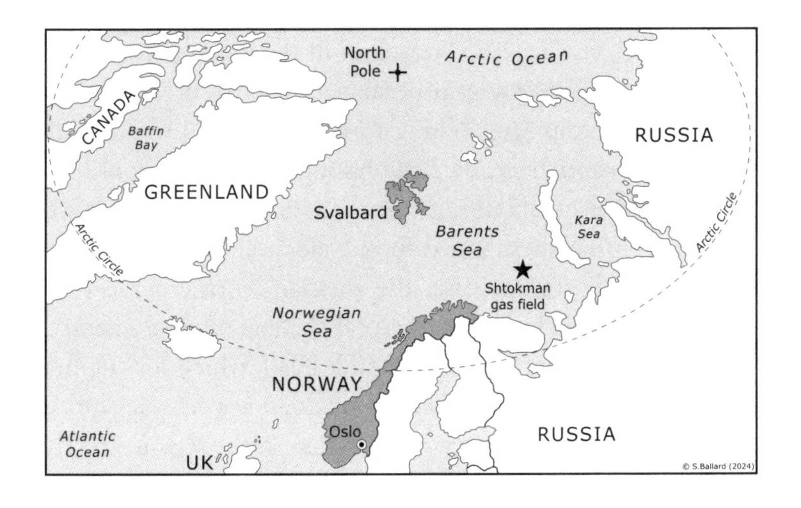

It is 2031. China and Russia launch a full-scale hybrid and cyber-attack on Britain. In parallel, attacks take place on undersea gas and oil pipelines, wind farms are sabotaged, and sea-bed telecommunications cables cut. World War Three has begun.

The war begins in the Arctic. In 2015, a small Chinese three-ship flotilla had conducted its first freedom of navigation operation in the American Arctic. It was just the beginning. Despite the Russo-Ukraine War, between 2020 and 2030 China and Russia steadily extend their reach, presence, and grip over the Arctic through investment, exploitation, and a marked increase in the military presence of their respective armed forces. Crucially, both Beijing and Moscow first reject, and then in 2027 formally withdraw from, the Treaty of Spitsbergen which granted Norway sovereign rights over the strategically vital Svalbard Archipelago but also gave assured access to Russia and other Arctic States. By 2028, China is using its growing military presence in Murmansk to influence almost all the countries bordering the Arctic region. And then...

In 2030, China invites Russia to take part in the largest military exercise ever seen in the South China Sea—Exercise Great Wall of Steel 34. The 2025 "ceasefire" in the Russo-Ukraine War sees Moscow effectively gain permanent control of Crimea and Donbas and encourages Beijing to believe that NATO is little more than a paper tiger. By 2030, having learnt the lessons of its military failures in Ukraine, the Russians are also well on the way to reconstituting their armed forces, albeit at great cost to the Russian people. In any case, the Russian Northern Fleet has been little affected by the war beyond the loss of naval infantry.

The exercise coincides with Vostok 2030, which for the first time takes place in the Arctic. Vostok 2030 is a seismic shift in the size and scale of military mobilisation in the region, and the first major joint Sino-Russian exercise. Over 500,000 troops, 20,000 tanks and 1,000 aircraft take part in the month-long exercise, along with a sizeable chunk of the combined Russian and Chinese Northern Fleets, including surface and sub-surface assets, as well as conventional and nuclear capabilities. Vostok 2030 is coordinated by the newly built and highly sophisticated Joint Russia/China Arctic Command Centre in Murmansk. It begins in the High Arctic with both countries testing their now extensive integrated anti-access/area denial (A2/AD) capability, reinforced by a sudden and large concentration of forces in the Greenland-Iceland-UK-Norway Gap. During the deployment, Russia also announces it has decided to extend its control over the continental shelf in the Arctic, a claim that China immediately recognises as "in line with the spirit and letter of international law". Not surprisingly, Russia also announces that China's "historic" claim to the South China Sea is legitimate.

In early 2031, Exercise Tsentr 31 follows fast on the heels of Vostok 2030 with the aggressive deployment of military capabilities, all of which are clearly designed to deter any Western response by moving sizeable forces back into the Arctic region. As it begins, another massive Chinese military exercise begins in the South China Sea that threatens Taiwan and possibly Japan, forcing the US to reinforce the Indo-Pacific theatre in great strength. In any case, US domestic politics are a mess and there is little support for Americans to get involved in what looks to many like yet another European conflict.

Tsentr 31 seems like another of a series of joint Sino-Russian exercises to test the ability of a closely coordinated force to operate—although ominously for the first time includes both Russian and Chinese submarine-based nuclear deterrents alongside advanced Russian and Chinese land-based nuclear missile systems. The exercise signals the development of a significant Sino-Russian Arctic-specific strategic deterrent with precision strike capabilities that if deployed could penetrate Western early warning systems undetected. During the first half of 2031 there is a steady build-up of Chinese and Russian forces as Tsentr 31 continues. In March 2031, Russia declares its intention to contest Norway's sovereignty over the Svalbard Archipelago and announces in the Russian media that "Svalbard is henceforth to be regarded as part of the Motherland." China and Russia also inform the Arctic Council of their joint intention to widen their exploration for resources in the uncontested region of the Arctic.

In August, Russia demonstrates its readiness and force projection by quickly expanding its military footprint, deploying three more brigades, 300 tanks, artillery, and surface-to-air missiles to the west of Murmansk, close to the Storskog crossing on the Russo-Norwegian border. The deployment clearly implies a threat to Norway's North Cape, which commands the ingress and egress of the Russian Northern Fleet from Murmansk. It also implies an attack on the territory of a NATO ally that could invoke Article 5 and collective defence. The deployment takes NATO planners by surprise, having been undertaken within a three-week period in August (when much of Europe is on the beaches); it suggests that the old Gerasimov Plan to move large-scale forces rapidly to anywhere on Russia's periphery is now reality. Moscow does not pull its punches. The Russian media states repeatedly that the deployment is a clear demonstration to NATO of Russia's military might in the region and its determination to secure what is called "Russia's critical and legitimate interests" by all possible means. Moscow also cites the long-term threat posed by Finland and Sweden joining NATO and what the Kremlin calls the upgrading of the Alliance's deterrence and defence plan, which had been completed in 2030.

In September, the US, Britain, and Norway agree that Russia's actions are a clear indicator that Moscow intends to increase its control over the Arctic region. NATO forces operating in the Atlantic and Mediterranean are thus directed to move northwards toward the Arctic. Crucially, neither British nor Norwegian intelligence believe North Cape or the wider Finnmark region is the target, but rather the Svalbard Islands, which explains Russia's expansion and deployment of A2/AD capabilities and long-range strike forces. For Russia, control of Svalbard would not only increase the Russian A2/AD "bubble" over the Arctic, but also allow for increased long-range strike capability. This would effectively slam the door shut on NATO's ability to move forces through the region or threaten offensive action against Russia via the high north. It would also have profound implications for the defence of Europe as it would threaten the ability of US forces to securely move across the Atlantic and reinforce Europe rapidly in an emergency.

On 12 September, the Russian military suddenly and rapidly reinforces deployed advanced A2/AD systems on Franz-Josef Land and brings them up to full operating capability. Russia's anti-access umbrella now stretches south and covers the whole of the Barents Sea. Chinese "scientists" on the southern tip of Svalbard and in ports on the east coast of Greenland are then "activated" by Beijing and begin to spread out across the landscape. Their main mission is to monitor (but not engage) a small Danish Special Forces group on routine patrol in Greenland's vast wilderness. They are subsequently reported by US intelligence to be elements of the People's Liberation Army Special Operations Force.

Two weeks later, on 25 September, Russia boldly declares the operational readiness of a new Arctic A2/AD system of systems. Russia 1 (Moscow's premier TV channel) declares that the aim of the system is "a Russian gift to the Arctic designed to ensure peace, stability, security, but above all prosperity" for the people of the region. The declaration draws immediate international condemnation, particularly in the wake of the news that there are Chinese military personnel on Svalbard. An urgent meeting of the UN Security Council is convened by the US, Britain and France to discuss this blatant contravention of the Spitsbergen Treaty and to demand the removal of Chinese forces on Svalbard. The three Western Permanent Members also propose a UN resolution condemning the militarisation of the Arctic by China and Russia. Beijing and Moscow veto the motion. On 27 September, Russian Special Forces begin a clandestine infiltration of Svalbard to link up with Chinese forces.

The Icelandic government, believing it is threatened by Sino-Russian actions, refuses to protest Chinese or Russian actions in the region for fear of what might be coming, and because both Reykjavík and the Greenland "capital" Nuuk, are awash with Chinese money. Nuuk also refuses to condemn Russian action which leads to profound political divisions in Copenhagen over how best to respond.

In the wake of an urgent meeting of the North Atlantic Council on 2 October, Littoral Response Group North (LRG (N)) is activated and ordered to deploy northward and take up position in the Greenland-Iceland-United Kingdom-Norway Gap. Two Strike Companies from 45 Commando, Royal Marines, and a detachment from 1 Marine Combat Group, Koninklijke Mariniers (Netherlands Marine Corps) embark their respective amphibious assault ships. The 1 Star force headquarters is flown out to join the task force as the force prepares to move. A third strike company from 45 Commando, Royal Marines, which is exercising with US Marines and Norwegian forces as part of the Teamwork 31 exercise, is withdrawn from Bardufoss and rapidly establishes a Company HQ and defensive positions in the vicinity of the Norwegian Arctic city of Kirkenes to block any Russian incursion towards North Cape. LRG (S), operating east of Suez, is immediately ordered back to Britain, whilst on 3 October, LRG (N) is directed by UK Navy Command to prepare to conduct operations against Russian and Chinese forces in the vicinity of Svalbard to secure access to the islands for NATO follow-on forces.

On 6 October, Norwegian fishing vessels, supported by the Norwegian frigate *HNoMS Roald Amundsen*, suddenly encounter multiple Chinese and Russian fishing vessels northeast of Svalbard, operating in the Norwegian Economic Exclusion Zone (EEZ). The *Roald Amundsen* warns the fishing vessels to stand-off from Norwegian naval vessels and to move out of the EEZ, but the Chinese and Russian fishing vessels respond by making a series of uncharacteristically "reckless" passes close to the *Roald Amundsen*, forcing her to veer away to avoid a collision. Russian military aircraft also breach Svalbard airspace several times and over-fly the *Amundsen* as it attempts to disperse the Russian and Chinese fishing fleets, and repeatedly penetrate Norwegian air space near North Cape. The *Admiral Amelko*, an *Admiral Gorshkov*-class Russian frigate, together with a Type 054A Chinese frigate, detach from the Northern Fleet and sail at flank speed towards Svalbard to provide "protection" for Russian and Chinese fishing vessels, which Moscow claims are under attack. Following a series of warnings, the *Amundsen* intercepts, and boards the Russian fishing vessel *Arina*.

On 7 October, widespread reports appear in the Russian media speaking of aggressive and "disrespectful" actions by the Royal Norwegian Navy against Russian fishing and trade vessels operating in international waters. Social media accounts across multiple platforms explode with comments and stories "highlighting" Norwegian recklessness and aggression, and accuse Oslo of being latter-day Vikings unnecessarily escalating tensions in the region. Stories also begin to appear suggesting that NATO is determined not only to infringe on Russian commerce, but to create a portal from which the Alliance can control a future Northern Sea Passage between Asia and Europe that could open due to melting Arctic ice.

On 8 October, following the boarding of *Arina* there are unconfirmed reports of an explosion on board the *Admiral Amelko*. Russian media and social media immediately report "an attack" on the ship from a "NATO military force on Svalbard against Russian naval forces legitimately protecting the Russian fishing fleet operating in international waters." It is information warfare—fake news. The real warfare is just about to begin.

D-Day in the Arctic

9 October 2031, two flotillas of frigates and destroyers from both the Russian and Chinese fleets break off from Exercise Tsentr 31 and move to Svalbard to "assist in the rescue and recovery of the *Admiral Amelko* and to establish a screen to protect the fishing vessels." Crucially, they carry the latest Tsirkon hypersonic anti-ship missile. Two Russian *Laika*-class nuclear attack submarines also head towards the Norwegian coastline to prevent the Norwegian Navy from deploying into the region. A Russian carrier task group and amphibious forces also move south, detach near Svalbard, and prepare for an amphibious landing.

On 10 October, both Russia and China accuse Norway of destabilizing the region and acting aggressively, whilst Russia blames Norway for the sinking of the *Admiral Amelko*. There is widespread outrage in Russian media even though Washington reveals satellite photos of the *Amelko* at anchor at the Severomorsk fleet base. China also condemns Norwegian actions and cautions NATO following the swift activation of UK and Dutch naval and amphibious forces.

On 11 October, LRG (N) begins to move into position several hundred kilometres south of Svalbard, as it prepares to insert teams onto the islands to link-up with the Norwegian military. The North Atlantic Council, meeting in emergency session, demands that Russia and China step down their forces. The NATO hope is that the arrival of the UK-led amphibious force will be enough to deter any further aggressive acts by China and Russia and de-escalate tensions. The UN and EU also seek to calm the situation and separately propose an international-led investigation into the explosion on the *Admiral Amelko*. The Alliance also hopes that by establishing a presence off the coast of Svalbard it will enable a stronger NATO force to be deployed to the region.

However, as the crisis intensifies, Chinese and Russian cyber and electronic warfare units also begin to target key infrastructure and communications capabilities in Svalbard and on mainland Norway. They also "blind" US, British and French surveillance satellites, drastically reducing any notice of impending Russian and Chinese actions. At the North Atlantic Council Germany also questions the deployment of a major NATO combat force, stating that Berlin's lawyers do not believe an attack on Svalbard would trigger an Article 5 contingency. Subsequently, Germany refuses to permit any of its forces to participate in NATO operations.

On 12 October, the aircraft carrier *HMS Queen Elizabeth*, the core capability of the UK's Carrier Strike Group, departs Portsmouth ahead of schedule and moves northward at speed towards the Arctic region. The Dutch amphibious ship *HNLMS (His Netherlands Majesty's Ship) Rotterdam*, with a company of marines aboard and a contingent of escort ships, departs Den Helder and is directed to link-up with the *Queen Elizabeth* Carrier Strike Group. The US also increases aerial surveillance in the Arctic region, whilst a US Marine Expeditionary Unit currently operating in the Mediterranean Sea is redirected north towards the Norwegian Sea. Unfortunately, having prematurely scrapped *HMS Albion* and *HMS Bulwark*, the Royal Navy's two amphibious assault ships, the Royal Marines must seek a "lift" on the *Rotterdam* with *RFA Largs Bay* laid up for a refit. Much of their equipment is left behind. They cannot operate from the *Queen Elizabeth* because she has embarked her full complement and London deems it too risky to conduct ship-to-objective manoeuvre (STOM) by bringing her close inshore.

Satellite coverage over the Arctic region remains degraded, whilst air-breathing systems offer some compensation there is a significant reduction in intelligence, surveillance, and reconnaissance flows. China also, and suddenly, unveils a hitherto undetected and complex web of A2/AD systems along the Greenland

Coast and the southern tip of Svalbard. Chinese forces on Svalbard also reveal themselves and begin to deploy mobile advanced anti-ship ballistic and cruise missiles on the south of the island, even though the US also manages to deny Russia and China their own satellite surveillance. Both sides are pretty much blind as the crisis intensifies, although Russia and China continue to use smart sensors and aerial surveillance to maintain what the Americans call "the intelligence grab." Russia also declares an air-defence identification zone over the whole of the Arctic and begins to challenge the aircraft of all other countries operating in the region.

13 October 2031

And then it starts. Early on 13 October, key early-warning systems on Svalbard are attacked with direct fire from a swarm of drones. The Norwegian Navy responds by deploying the frigates *Otto Sverdrup* and *Thor Heyerdahl* to southern Svalbard. Russian intelligence reports the Norwegian 1st Submarine Squadron is also preparing to deploy the super-quiet *Ula*, *Utsira* and *Utstein* boats, which could pose a significant threat to the amphibious force. Russia responds by activating several brigades, tanks, artillery, and surface-to-air missile capability to the west of Murmansk, close to the Storskog border crossing, but stops just short of entering Norwegian territory. They then dig in.

0001 hours: A Russian nuclear attack submarine detects the Norwegian flotilla moving towards Svalbard and is ordered by the Russian Arctic Command to strike and sink the *Otto Sverdrup*.

0100 hours: Multiple cruise missiles strike facilities in Bardufoss, where US Marines, Norwegian military, and British Royal Marine Commandos have been conducting tri-lateral training. The initial assessment is catastrophic, with many vehicles and a lot of equipment destroyed. The high number of personnel on the base were preparing for follow-on movement to Kirkenes, resulting in a high number of casualties. Russian forces near the Storskog border crossing are also put on high alert and ordered to prepare a sweep through northern Norway along the Kirkenes-Narvik axis. Finnish and Swedish forces deploy in support of their Total Defence doctrine but lack significant expeditionary forces.

0110 hours: Chinese and Russian military action against Svalbard begins and sustained air-strikes quickly immobilise the limited local defences on the island.

0300 hours: Russian and Chinese naval infantry advance forces take control of Svalbard's outlying island and then swiftly move to the main island and seize key coastal infrastructure and port facilities. More A2/AD systems are offloaded by close-in shipping and move away from debarkation points. Chinese and Russian Special Forces secure the southern peninsula and disperse to defend the A2/AD systems. Meanwhile, hundreds of Russian paratroopers appear over Svalbard's coastline charged with quickly securing ground and establishing beachheads so that additional A2/AD and other missile systems can be disembarked.

0300 hours: Britain goes dark as much of its undersea telecommunications network is destroyed together with gas and oil pipelines from Norway. Most Britons are asleep, but London immediately declares a major emergency even while much of its unprotected critical infrastructure, including hospitals and police, are paralysed by extensive cyber-attacks. Very rapidly what passes for civil defence collapses even before the population wakes up.

0315 hours: As soon as the force commander receives notification that the Russian military has invaded, LRG (N) begins to move with the amphibious group poised only some ninety kilometres to the southwest of Svalbard. However, intelligence reports also reveal the presence of several S-500 missile systems on the south of the island with additional military capability being offloaded. Russian Special Forces (Spetsnaz) are also reported to have seized critical infrastructure, whilst US intelligence reports confirm an extensive A2/AD "system of systems" is already in place along the southern Svalbard coastline and linked to similar systems established on the east coast of Greenland and on Franz Josef Land.

0325 hours: The threat of anti-ship cruise missile attacks becomes pressing and makes the immediate LRG (N) position untenable, forcing the Allied fleet to move further out to avoid the missile threat. The window to deploy any strike teams suddenly becomes much more limited and continues to reduce. Worse, despite the attack on Bardufoss, several NATO allies with no forces in the fight refuse to trigger an Article 5 contingency.

0330 hours: With NATO divided and its follow-on forces out of range the Alliance is unable to support British, Dutch and Norwegian forces in the region or have any impact on the events taking place on Svalbard. The threat posed by the sudden and massive increase in A2/AD capabilities demands a much heavier Allied force if the Chinese and Russians are to be dislodged. Despite that, it is decided to employ advance forces to find and strike Chinese and Russian high-value targets (HVTs) on Svalbard in order to open a corridor for a rapidly assembling US force.

0450 hours: British and Dutch Commandos, on "strip alert," are deployed to Svalbard, with several "Strike Teams" conducting a hazardous 90km ship-to-shore movement to land on the southwest and northwest of Svalbard.

0630 hours: Covert insertion of the first wave of multiple teams is successful. Insertion of the second wave by air is aborted following indications that aircraft were being tracked and targeted by Chinese anti-air capabilities.

0700 hours: Multiple strike teams conduct infiltration from the coast of Svalbard into the interior. Some move to link up with Norwegian forces, whilst others begin to find and destroy HVTs, primarily A2/AD systems. This first action is considered essential to allow follow-on forces into the region. The intent is to achieve rapid success, supress Chinese and Russian A2/AD, and allow the arrival of US forces to serve as a catalyst for so-called off-ramp negotiations.

0845 hours: Following their covert insertion, the strike teams quickly find and strike HVTs on the island using their strategic capabilities and drone-delivered munitions from distance. However, Russian and Chinese Special Forces on Svalbard continue to conceal and fortify deployed A2/AD systems.

0900 hours: Chinese and Russian forces from their respective carrier groups and amphibious assault ships launch a massive maritime-amphibious operation on Svalbard supported by air power and protected by the A2/AD systems.

0915 hours: Russia launches a series of "hunter-killer" operations using helicopters in conjunction with ground forces to find and destroy British and Dutch strike teams operating on the islands.

1015 hours: Russian find-and-destroy operations prove to be highly effective, with four British strike teams rapidly immobilised or eliminated. The operation continues with further "hunter-killer" aircraft deployed to locate and strike NATO forces on Svalbard. Allied attacks on A2/AD systems are halted.

1400 hours: The *HMS Queen Elizabeth* Carrier Strike Group (CSG) remains at standoff distance from what is believed to be new Russian A2/AD and the wide weapons engagement zones it has now established. Consequently, close air support for forward deployed forces is significantly limited.

1500 hours: LRG (N) merges with the *HMS Queen Elizabeth* Carrier-Strike Group to form a hybrid Expeditionary Strike Force (minus LRG (S), which is still on its way back from Suez). Unfortunately, LRG (N) can only offer limited logistical support to strike teams on the island and has little capacity to redeploy forces due to the A2/AD threat.

1700 hours: NATO adapts. US-led air power operating from bases in Britain, supported by the F35B Lightning 2 strike aircraft on *HMS Queen Elizabeth*, strike Chinese and Russian A2/AD capabilities on Svalbard. Over seventy Chinese and Russian personnel are killed, and several mobile A2/AD systems destroyed, but six Allied aircraft are shot down with the loss of their pilots.

1900 hours: Despite Allied action to prevent the occupation of Svalbard, the "rapid dominance" and "shock and awe" tactics of Chinese and Russian forces have resulted in the swift and widespread seizure of both North and South Svalbard islands, with over 20,000 troops landed during the day and more following every hour. S-500 anti-aircraft missile systems with an operational range of 600 km are rapidly off-loaded and moved to defensive positions throughout the archipelago. The layered and integrated web of Chinese and Russian A2/AD systems now stretches from Greenland through Svalbard to Franz Josef Land, blocking all air and sea approaches to the Arctic and threatening to impose a disproportionate cost on any NATO follow-on forces. The Chinese and Russian carrier groups are held in reserve in the South Barents Sea, with close air protection covering the offload on Svalbard, as well as protecting vital ground forces. D-Day in the Arctic is effectively over.

Britain and Norway are in chaos. Only Finland and Sweden with their Total Defence Concepts applied to the full have beaten off Russian cyber and information warfare. The only good news is that the information warfare the Chinese and Russians had launched in parallel with its cyber-attacks failed simply because Moscow and Beijing had not realised how fragile Britain's critical infrastructure was, despite repeated assurances by several governments that they had taken steps to prevent just such an attack. They had not. Communications have collapsed and vital information to the public blocked.

October 15: The Allied follow-on force arrives and prepares for strikes and, if ordered, forcible entry onto Svalbard. However, the swift action of Chinese and Russian forces to close-off access to the Arctic has worked and Allied commanders quickly realise that any military action to take back the initiative would be costly and could easily escalate. The Chinese and Russian anti-access threat is proving highly effective. Commando strike teams are successful in locating and targeting HVTs, but not to the extent that would allow the US Navy and Royal Navy to close in on Svalbard in sufficient strength, or to allow effective targeting of the now established integrated air defence system by the US Air Force. In any case, the main body of US strength is in the Indo-Pacific. NATO is divided, the allies have been caught off guard, and significant offensive action will now be required to remove Chinese and Russian forces from Svalbard, which given the dangers of escalation few of the European allies are willing to countenance because they do not want to end up like Britain. As food supply chains begin to break down and other essential services collapse, Britain's already fractious and divided cities start to witness social disorder as "communities" begin to self-organise led by "community leaders," often fighting against each other. Law and order steadily retreat in the face of mass looting and violence.

Having seized Svalbard to "protect the Arctic from NATO aggression," Russia and China now seek to de-escalate tensions in the Arctic before the situation spirals out of control. NATO is shell-shocked and Allied leaders and commanders recognise that China and Russia have executed a near flawless *fait accompli* to bar the northern gate to the Arctic. Russia and China have the military capability and troops in place to dominate the region by force. The Sino-Russian partnership has, by the sheer scale of its presence, secured critical sea lines of communications and control over the Arctic's vast resources. It has also gained control of the access to the Bering Sea, greatly enhancing the ability of Russian nuclear ballistic missile submarines to operate undisturbed in their bastions and extend uncontested Sino-Russian influence far out into the North Atlantic. The NATO defensive line that stretches from Greenland to Iceland and then onto the UK is outflanked, preventing reinforcements from across the Atlantic reaching Europe. It also enables both China and Russia to threaten vital sea-lines of communication far more easily than hitherto.

Taking Svalbard was a risk for both China and Russia, but the strategic gains have been enormous. For NATO and Europe, the impact is quite the reverse: humiliation, division, and effective expulsion from the Arctic, and a crippling blow to Western cohesion. For all that, the US and its European allies have no option but to cede control of the region to China and Russia or risk a catastrophic outcome. And Britain? It is a broken mess.

And then, deep in the South China Sea, the *USS Ronald Reagan*, and its task group...

With the Americans busy elsewhere and Britain effectively paralysed, suddenly a British Prime Minister is faced with the appalling prospect of reaching for the nuclear button far earlier than she had ever feared. And all because for years London had been playing that most dangerous of political games—defence strategic pretence!

Lessons

The Chinese and Russian attack succeeds because whilst there had been much political talk, as there always is in Britain, about an integrated defence across intelligence, cyber and kinetic security, reinforced by government assurances about enhancing resilience of critical structures, infrastructures, and people, little had been done. Britain was unable to absorb the strike and was thus unable to strike back at a vulnerable Russia in the first systemic cyberwar. London was effectively paralysed and out of the war before a shot was fired. Europe's unsinkable aircraft carrier had effectively been sunk and the ability of the US and Canada, not to mention Britain itself, to reinforce NATO's Forward Defence destroyed. Even if NATO had prevailed the Alliance had little ability to move forces quickly and securely across Europe due to a host of legal and physical barriers preventing military mobility.

Having rebuilt its armed forces since the debacle of the Russo-Ukraine War with the help of China, Russia had struck at the very place and in the very manner Britain was unable to defend. In the wake of Russia's February 2022 invasion of Ukraine, it had appeared London would respond to the threat but, as ever, the politicians had talked, and talked, invested a bit here and there and written a host of so-called integrated reviews. However, London's enduring obsession with the balance sheet simply masked its endemic curses—the confusion of interests with values, the short-term being favoured at the expense of the long-term, and the political pretending to be the strategic. There was always either no money or something else to spend money on.

For many years, Britain's leaders had been more concerned about hiding their defence unpreparedness than defending Britain. The consequence was deeply rooted defence strategic pretence. Britain's shiny new Navy was little more than a hollowed-out Potemkin force, whilst the British Army was so small it could contribute very little to the defence of Britain, let alone the defence of Europe, and the Royal Air Force had lost most of its experienced combat pilots, many of whom had left in disgust at the growing obsession of high command with political correctness.

A significant part of Britain's expenditure on defence had also been spent on non-military "rent-seekers," such as the National Cyber Security Centre and the *Dreadnought*-class nuclear-powered ballistic missile submarines, rather than actual fighting power. Worse, an obsession in government with diversity over unity had broken the link between the armed forces and much of Britain's population, making recruitment ever harder than it should have been. Patriotism had become a dirty word among the London *bien pensant* elites who "ran" the country. Britain was not alone in paying the price of such weakness, but it was meant to be a leading power in NATO. Europe and London's strategic pretence had profound consequences for the Alliance. NATO security assurances, even its much-vaunted Article 5 collective defence "guarantee," had been revealed to be a bluff. The Alliance quickly collapsed. There was always someone else who was going to defend Europe, but that someone else, the Americans, could simply no longer afford to carry Britain and the rest of Europe in the way it had since 1949.

Ultimately, Britain's defeat and the defenestration of NATO was not only due to a retreat from strategy but a retreat from realism. Without realism, there can be no strategy. It was a retreat that began in the dark days of 2010 when the gap between the ends, ways and means of security and defence grew dangerously wide in the aftermath of the financial and banking crisis. Despite the many British National Security Strategies, Strategic Security and Defence Reviews and so-called Integrated Reviews, which were meant to make Britain stronger, the opposite happened simply because security and defence policy was smoke and mirrors and worth little more than the paper they were written on.

INTRODUCTION

Systemic competition: the intensification of competition between states and with non-state actors, manifested in a growing contest over international rules and norms; the formation of competing geopolitical and economic blocs of influence and values that cut across our security, economy and the institutions that underpin our way of life; the deliberate targeting of the vulnerabilities within democratic systems by authoritarian states and malign actors; and the testing of the boundary between war and peace, as states use a growing range of instruments to undermine and coerce others.

"Global Britain in a Competitive Age: the Integrated Review of Security, Defence, Development and Foreign Policy,"
Her Majesty's Government, London, 2021

Virtue Imperialism

There is no such thing as a free Chinese lunch. In 2023, it was revealed that the Foreign, Commonwealth and Development Office (FCDO) was in advanced talks to hand over sovereignty of the strategically vital Indian Ocean island Diego Garcia to China-aligned Mauritius some 2152 kilometres or 1337 miles distant. It was a deal that reeked of the virtue imperialism that is doing so much damage to Britain and its interests. In December 2023,

London withdrew from the talks but the fact that they took place at all reveals a culture of grand strategic illiteracy in London. Virtue imperialism is a declaratory end supported by neither means nor ways, in the hope that the world will follow London's example by sacrificing its own vital interests for the sake of some ill-defined set of values. That is not how power works, nor how it has ever worked.

London's addiction to virtue imperialism is both strategically and geopolitically perverse, driven by the historic guilt of a small but powerful elite in London. Paradoxically, virtue imperialism has become the last preserve and vestige of British imperialism. Misplaced guilt seems to be a primary driver of British foreign and security policy at the upper levels of Britain's political and bureaucratic establishment, with profound implications for defence. Guilt about who the British are and who the British once were.

Diego Garcia is simply a case in point. Former Prime Minister Boris Johnson warned that any such transfer would be a "colossal mistake." He is right. Diego Garcia may be a tiny speck of an island in the Indian Ocean, but it also hosts the most strategically important British-owned US air and logistics base in the Indian Ocean. The FCDO response to the revelation it was engaged in such talks was to equivocate with the usual bromide that no decision was imminent and were Diego Garcia ever to be handed to the Mauritians the American base would not be threatened. Really? There can be little doubt the Chinese were pushing Mauritius to claim Diego Garcia and that China would love at some point to turn Diego Garcia into another of its ever-extending Blue Water Naval Strategy.[1] There are forty-seven official Chinese development finance projects in Mauritius, which is mired in China's Belt and Road Initiative.

The Integrated Review Refresh 2023 was sub-titled "Responding to a more contested and volatile world." The problem is that much of Whitehall, with the FCDO to the fore, simply does not believe in strategic competition or Britain's need to be strategically competitive. For the FCDO and their fellow travellers, if Britain should lead at all it should be through virtue. The problem is that no-one else is listening and by so often placing values before interests the contemporary political and bureaucratic culture in London undermines Britain's core security interests, and in this instance, those of the United States. By even agreeing the principle of a Mauritian claim over Diego Garcia they put at risk other British Overseas Territories. There are strikingly similar historical parallels between the claim by Mauritius for Diego Garcia (2152 km/1337 miles shortest distance) and the Argentine claim on the Falkland Islands (393 km/244 miles shortest distance). Precedent matters in the sovereignty game especially for lawyers for whom upholding increasingly pervasive international law is the be all and end all of policy.

The Mauritian claim is dubious at best. When in 1810 the British seized Diego Garcia from the French there were no permanent settlements in Mauritius and there never had been. The country of Mauritius simply did not exist. The Chagos Archipelago was made part of the British colony of Mauritius simply for administrative convenience. If there is any legitimate grievance to be had it is on the part of the descendants of those who were living on Diego Garcia at the time the joint UK/US air base was established between 1968 and 1973 and who were forcibly expelled.

The Diego Garcia saga fits a wider pattern of contemporary British foreign policy. *The Oxford Concise Dictionary* defines "imperialism" as "acquiring colonies and dependencies, or extending a country's influence through trade, diplomacy, etc." Thankfully, the age of Britain acquiring territories is over, but not so for China, which is why Integrated Review 2021 described "Global Britain" as being "in a Competitive Age." And yet London seems to be doing everything in its power *not* to compete in what is a self-evidently contested age. It is as though many in London no longer believe Britain has a right to pursue its legitimate critical interests. If those ruling a country do not believe in its rights to act as a country in an anarchic international system, that country is inevitably doomed to decline and fail.

Such virtue imperialism also signals London's retreat from realism and as such a retreat from strategy. Without realism there can be no strategy. The tragedy is that most Britons do not feel guilty about Britain's past—far from it—and in any case that was then, and this is now. Worse, Britain really does live in a strategically competitive world because since leaving the EU Britain is again facing such reality red in tooth and claw. For Britain to remain a wealthy and relatively powerful country, the need to compete effectively across the four great strands of grand strategy—diplomatic, informational, military, and economic—has never been greater.

The core contention of this book is that Britain's retreat from strategy is in danger of putting the British people in an increasingly dangerous position, caused by an elite Establishment in London that is not only risk averse, but power averse. Britain is also in crisis about what role the country should aspire to play in the twenty-first century. It is a crisis the British people instinctively sense, with its epicentre at the heart of government. The British people are also beginning to sense just how badly they are being failed by their leaders and how vulnerable they are fast becoming, even if they do not understand why. Consequently, the British today live with greater risk than for decades because their leaders too often reject the critical relationship between Britain's legitimate vital interests and the power and strategy needed to secure them. In a world in which realpolitik has made an unwelcome return, British weakness not only makes Britain less secure, but also its allies and partners.

INTRODUCTION

Potemkin Britain

London is bereft of strategic vision with leaders who have lost confidence in their own capacity to lead, which in turn has led to a small-power mindset and a weakened strategic culture. This has enabled other powers—friends, allies, and adversaries alike—to take advantage of Britain's exaggerated and self-imposed weakness.

In 1787, Count Grigory Potemkin is reputed to have constructed a mobile, fake floating village in conquered Crimea. The purpose was to impress the Empress Catherine the Great and visiting foreign dignitaries of the benefits of Russian occupation and mask the slaughter that had taken place of the Tatar people. Over the centuries the term Potemkin Village has come to describe any fake structure that appears to the distant observer to be far more substantive than it is. John Lewis Gaddis famously accused the Cold War Soviet Army of "Potemkinism," creating just enough capacity to imply far more capability.[2] Britain and its armed forces are equally guilty of Potemkinism today with profound implications for British strategy.

British grand strategy should concern the generation, organisation, and application of still immense national means in pursuit of critical national interests. Grand strategy is not military strategy. Indeed, successful grand strategy *never* defaults to military strategy: the latter is simply a servant of the former and just one of several instruments of national power. Military strategy is about matching military ends, ways and means. Grand strategy, as its name suggests, is about securing many more and far bigger ends, via much broader civil and military ways, through the application of immense means in pursuit of constant geopolitical objectives deemed vital to the country.

In London, what is presented as "strategy" is in fact tactics. Failed campaigns in Iraq and Afghanistan demonstrated such a failure, since they were not built on a core strategy with clear political ends. Rather, they were a composite of operational actions and open-ended tactics to no strategic end over a long time and at great distance over a very large space. In other words, Britain and her allies muddled along in an unforgiving sea (or perhaps desert) of strategic mediocrity.

The danger of Britain's Potemkinism was evident in a July 2023 report by the House of Commons Intelligence and Security Committee on China, which stated that the government's

> focus has been dominated by short-term or acute threats. It has consistently failed to think long term—unlike China—and China has historically been able to take advantage of this. The government must adopt a longer-term planning cycle with regards to the future security of the UK if it is to face Chinese ambitions, which are not reset every political cycle. This will mean adopting cross-government policies which may well take years to stand up and require multi-year spending commitments. This is something that will likely require Opposition support—but the danger posed by doing too little too late in this area is too significant to play politics with. For a long-term strategy on China—thinking ten, fifteen, twenty years ahead—the government needs to plan for it and commit to it now: the UK is severely handicapped by the short-termist approach currently being taken.[3]

Yvette Cooper, Labour's Shadow Home Secretary, went further when she warned about China's "all of state" offensive against the UK:

> We can't sort of drift ... It is so serious to our national security, not just today, but for many years into the future ... the government is failing to get ahead of the economic risks to our domestic security, which has meant that our response to a fast-changing security landscape has often been disjointed, disorganised and delayed ... We've been too slow to identify where risks lie, and where to safeguard national infrastructure.[4]

Despite all such failures over the past twenty years—and failed Britain has—London should still be capable of thinking and acting grand-strategically. As former Secretary of State Des Browne writes, "we need both ... the ability to think ... and crucially, act ... grand-strategically in the national interest. Doing so would be tantamount to a grand, distinctively British strategy whether or not it was labelled as such."[5]

Sun Tzu famously said that "the acme of skill is to defeat one's enemy without firing a shot." Such "acme" is about power and its application: British power in the here, now, and tomorrow, not the then and has-been. There are two critical questions to be addressed. First, why has British influence declined? Second, what role should Britain aspire to play in the world? After all, Britain remains a big power with a big economy reinforced by a whole raft of diplomatic, informational, military, and economic instruments of power. The need is pressing. Britain remains one of the world's leading economies and military powers, even if these days it is a decidedly regional-strategic Euro-Atlantic power. Ultimately, strategy is about choices and the more choices one needs to make to balance the ends, ways and means of the national interest the more informed such choices need to be. That means big clear thinking about big issues and a much better understanding of how to achieve Britain's legitimate goals and secure its critical and vital interests within a methodological framework for crafting both policy and strategy.

Levels of Strategy

Des Browne also writes, "The failure is not in the diagnostic exercise represented by Integrated Reviews but in the government's inability to devise policy of sufficient scope (both in thematic and financial terms) to engage the challenges they identify."[6] There are three levels of strategy central to this book:

grand strategy, national strategy, defence strategy (and by extension military strategy). However, a distinction must first be made between policy and strategy. Policy sets the overarching aims of a state across government. Strategy takes policy as its aims and objectives and sets out to achieve them.

Grand strategy is the application of all national means by government in pursuit of high geopolitical ends. Most obvious during wartime, grand strategy is also implicit during peacetime, even if unwritten. Grand strategy applies all diplomatic, informational (political declaratory), military and economic instruments of power. What makes grand strategy distinctive is that it should be founded on enduring principles and fundamentals of a state's power and interests and is thus less prone to short-term political considerations. Consequently, both the true interests and the relative power of a state are evident in grand strategy.

National strategy sets out in writing the codified strategic and political goals of a government at any one time, both domestic and foreign. National strategy is the vehicle for policy over the short and medium term, and should be the best available plan to realise considered ends via suitable ways and necessary means. It should thus be entirely consistent with a state's grand strategy. National security strategy is a subset of national strategy, in the same way security policy is the servant of national aims and defence policy a servant of security policy. Crucially, national security strategies identify and prioritise threats, risks, and the resilience a state needs to function at times of crisis and the method by which a state plans to act. National strategy also distinguishes, or at least it should, between critical and general interests, as well as between interests and values.

Defence strategy concerns the place, scope, role, and missions of the military instrument of power in the mix of instruments a state can bring to bear. The emphasis in defence strategy is on defence architecture, the military domains and increasingly the minimum military capabilities that a state needs across air, sea, land, cyber, space, information, and knowledge. Defence strategy should also be dynamic and able to identify the types of capabilities a state believes it will need over the medium-to-longer-term to maintain credible deterrence, effective defence, and sustained engagement, if needs be. Defence strategy is based on the assumptions and strategic judgement of national strategy, but its focus should be on the worst possible case as well as the most likely array of threats and contingencies, with defence planning assumptions to match.

Military strategy is where the rubber of grand, national and defence strategy hits the road of military reality. Much of this book will be spent considering this crossroads in British strategy precisely because it concerns the strategic application and utility of the military instrument of power in service of the supreme political authority of the British state. If the gap between Britain's ambitions, aspirations and intentions and the scale and scope of the military instrument of power is too wide the risk of engagements that catastrophically fail grows exponentially.

Power and Strategy

Power imposes responsibilities. Power and strategy are thus inexorably and intrinsically interlinked precisely because power is the "commodity" that strategy applies. Therefore, before any discussion of strategy can take place one must define state power: a state's ability to exercise influence over other states within the international system. Influence can be coercive, attractive, co-operative, or competitive. Mechanisms of influence can include the threat or use of force, economic interaction or pressure, diplomacy, and cultural exchange.

Power is also fungible and comes in several forms, both soft and hard. Soft power incorporates issues such as language and cultural reach. According to the Global Soft Power Index 2022 Britain is the world's second most influential soft power actor after the United States.[7] However, it is economic power that ultimately provides the bedrock for sustained comparison of power between states. According to the International Monetary Fund (IMF) Britain in 2023 had the world's sixth largest economy after the US, China, Japan, Germany, and India; although the same IMF data suggests Britain has the world's tenth largest economy in terms of purchasing power parity, with China before the US and even Russia above Britain.[8] The UN Human Development Index also has Britain ranked eighteenth, behind first-ranked Switzerland and second-ranked Norway, although Britain is ahead of the US and France. When it comes to hard military power, the core subject of this book, the much-respected Global Firepower Index ranks Britain sixth in the world after the US, Russia, China, India, and South Korea, but significantly ahead of France and Germany.[9]

Britain has an official population of some 68 million souls, which ranked it the twenty-first most populous country in the world in 2023, well behind India (1.43 billion) and China (1.42 billion). However, size of population is as much a curse as a blessing for a state, particularly for developing countries such as China and India.[10] Britain is also the world's seventy-eighth largest country by landmass according to the World Atlas, with Russia by far the largest country, which is a curse because Moscow must govern across eleven time zones covering two continents.[11] Britain is also a major trading nation. In the twelve months prior to May 2023 the British exported some £854.1 billon of goods and services whilst it imported £904.6 billion.[12] Any disruption to global trade is thus a threat to the British economy and society, which means strategic isolationism is not an option. In other words, Britain must engage in the world—even if some Britons would appear to think otherwise—and play a fulsome role in the protection of sea and air lines of communication and global patterns of trade.

Taken together across diplomatic, informational, military, and economic (DIME) instruments of power, Britain remains a genuine Great Power on paper at least, but it is no superpower. Britain is also a Goldilocks power—neither too big nor too small—and there is a lot a state can do if the area of its governance is compact, its society and economy well-developed, its institutions highly networked and its armed forces capable. The problem is that like several Western European states Britain is also caught in a power trap. Power is utterly unforgiving for those who have it because they cannot hide from the fact of it.

Unfortunately, for all of Britain's intrinsic strengths, British power is also to some extent illusory. Whilst still nominally the fifth or sixth biggest economy in the world, the manifold crises with which London has had to contend since 2008 have also left Britain heavily indebted with low economic growth and relatively high inflation. Britain has an ageing population with a falling GDP per capita not least due to the waves of poorly managed mass immigration successive British governments have encouraged, which have depressed wages and placed social structures and systems under intense pressure, most notably the hulking and inefficient National Health Service.

According to VisualCapitalist.com, Britain in 2023 was the world's twenty-fourth most indebted country by debt to GDP (100% debt to GDP), behind Japan which has a 257% debt to GDP ratio, whilst the US is on 133% and France 116%.[13] However, these figures are somewhat misleading as much of Japan's debt is "owned" by the Japanese people, who maintain a high level of savings compared with citizens in other advanced economies such as Britain, France, and the US. Tokyo is also relatively stable and because of that Japanese policy is less inhibited by debt, as evidenced by Japan's marked increase in defence investment. Current levels of British indebtedness act as a drain on the public investment upon which defence depends. And Britain's strength depends on the fact that, taken together, its various instruments of power afford London great influence if properly applied through a coherent strategy clearly focused on the British national interest. And therein lies Britain's essential problem: it lacks such application and strategy. There is a profound lack of strategic ambition and culture amongst the highest political and bureaucratic actors in London, allied to profound confusion over the value and utility of such power.

The American economist J. K. Galbraith famously said that "power is as power does."[14] If Britain is to return to strategy, London—its political leaders and the high bureaucracy that nominally serves it—will need to again grip Britain's fundamental interests and reacquire a grasp of twenty-first century realpolitik. Only then will Britain do what it can do and must do given the world in which it exists. At the very least, Britain must rediscover political realism and overcome the profound confusion of values with interests which so dilutes its ability to translate power into effect. This tension is typified by the confusion in London between what has been called an ethical foreign policy and the British national interest, which has for too long critically prescribed strategic analysis, profoundly undermined strategic clarity, and often rendered British "strategy" no strategy at all. Not everything a country must do to protect itself and prosper is about the political here-and-now, but that is precisely what many of Britain's political "leaders" seem to believe.

Furthermore, as President Bill Clinton said, "it's about the economy, stupid." It is economic growth that provides the basis for all forms of other "power," be it health, education, or any other activity of government in a democratic society. Economic power affords government choices, but to make those choices leaders must understand the scale, nature, and gravity of the choices to be made, not just their own political interests, and have criteria for so doing. Take Britain's energy policy, such as it exists, which reveals a form of post-imperial political hubris too often driven by a desire to save the world from itself even at Britain's own expense. Britain's commitment to Net Zero by 2050 allied to punitively high taxation and inflation is acting as a dampener on economic growth and thus Britain's ability to compete in the world. The Centre for Economic and Business Research (CEBR) has warned that the costs of decarbonisation will accelerate the shift in economic power away from the West towards China and others, whilst the Organisation for Economic Cooperation and Development (OECD) has warned that Net Zero will leave Britain's economy £60 billion smaller by 2050 and cost the world as much as $3.6 trillion (£2.8 trillion).[15]

This is compounded by the refusal by successive governments to exploit very significant natural resources available to the British, such as reserves of shale oil and gas as part of a meaningful and considered transitional strategy. This focus on values at very high cost also reveals a problem that runs throughout this book: the lack of hard thinking in London about the national interest and the strategy and actions required to secure Britain's longer-term prosperity and security. This is partially a consequence of the undue influence of pressure groups on government and the power growing legions of lawyers seem to have over successive administrations in an increasingly law-bound society. Much political energy is expended by government simply fending off such groups in the media and the courts whilst the British people look on. They also perceive a dishonesty in politicians of all stripes who promise big on a raft of issues important to them but fail, or rather refuse, to deliver. Over time popular frustration with the political class and structures is eroding London's ability to govern. It will only get worse.

The Retreat from Strategy

What is British strategy? On the face of it that would seem a silly question. The British government produces a plethora of "strategies" for just about everything. At the core of the problem is that a multitude of "strategies" do not together constitute a strategy at all but a form of anarchy within government, which is as much a retreat from reality as a retreat from strategy. The reasons for this situation are both vital and simple. One problem is consistency of leadership. As Des Browne writes,

> in the last thirteen years of Conservative-led government, we have seen seven Foreign Secretaries, seven Defence Secretaries, innumerable junior ministerial changes and, perhaps most crucially, five Prime Ministers. This turmoil, coupled with an enormously challenging geopolitical backdrop, has placed Britain at a significant disadvantage in engaging our strategic adversaries and pursuing the national interest.[16]

Britain is facing a host of emerging threats generated by ever more complex global frictions and yet London has lost the ability to make not just hard choices, but correct choices, by applying statecraft to policy to properly cohere them. The British have traditionally been good at painting the big strategic picture and offering long-term ambitious solutions. No more! Strategists are no longer found in government, where the emphasis is on the immediate, the political and the short-term (even if it is often dressed up as the long-term).

There is little new in the political dominating the strategic. It is simply that the friction between the two has become progressively more acute as the country that Britain is—a strange post-imperial, post-Europe global nowhere Britain—sits uncomfortably with the Britain that politicians too often pretend remains. To maintain the fragile appearance of power London has become utterly averse to any risk that would reveal the truth, even at the expense of strategic sense. Fearful of being "outed," the consequence is a form of appeasement of any external threat that might reveal the extent to which Britain, its institutions and its strategic culture are little more than paper tigers.

There are two essentially contradictory forces at work that reveal the increasingly threadbare nature of British grand strategy. First, a growing geopolitical threat, and with it the still distant but not impossible prospect of a Third World War, that demands of Britain not only a marked increase in security and defence investment, but also a new balance to be struck between power projection abroad and the protection of the British people at home. Second, an increasingly fragmented society in which much of the high political and bureaucratic Establishment (the mix of Westminster and Whitehall referred to in the book as London) seems to have abandoned pursuit of the national interest in favour of fashionable, globalist, *bien pensant* causes. Successive governments' inability to implement economic policies that generate the required long-term rates of growth is a key example. Grand strategy relies first and foremost on leadership-generated national unity.

The pre-requisite for sound strategy no longer exists in Britain, not least because so many of the British people regard their leaders with contempt. Such opprobrium is not entirely fair. Governance of a complex democracy in a social media age is an extremely difficult art. Abraham Lincoln might have reconsidered his truism if he lived today that one might be able to please all the people some of the time, but it is impossible to please all the people all the time. It is very hard for politicians to even please some of the people some of the time these days partly because expectations have become so bloated. Equally, there *is* a problem of government that compounds those challenges. It is not only a retreat from strategy from which Britain suffers, but too often bad strategy, much of it the result of a profound misunderstanding about the relationship between strategy and politics. Examples abound, from the HS2 high-speed rail project to healthcare to immigration to Net Zero policy, even to Britain's response to the 7/10 massacre of Israeli citizens by Hamas and Tel Aviv's response, and of course, Ukraine, to name but a few.

There are several causes of this malaise. First, reforms made since 1997 have set up alternative poles of power in Britain. The powerful administrative state and the equally powerful legal state are relatively recent phenomena that often compete for power with the political state. Second, the political state has been weakened by the devolution of powers to the constituent nations of the United Kingdom except for the largest, England. Years of rule-taking from Brussels has made the administrative class used to a certain sort of power, the day-to-day horse trading that is the EU, rather than the formulation of grand strategy. The obvious consequence of this friction is immigration policy. In 1997, the population of Britain was some 58 million people with services and housing designed to serve it. In 2023, the official population is now some 68 million people with the real figure probably over 70 million. And yet, political, bureaucratic, and legal inertia has not seen services or housing keep pace with population growth.

London also tends to "gold-plate" the rules of the rules-based international order in the belief that strong rules can be an alternative to power. An example is Rule 39 of the European Convention on Human Rights, which allows for a temporary injunction to prevent what might be a harmful act by a state, but which is not binding. The problem with such rules-based orders for Great Powers is precisely that they have rules and rules constrain action for a reason: to prevent extreme state behaviour. Extreme behaviour is the kind of military adventurism Putin unleashed on Ukraine. Only the Great Powers have the instruments of power to prevent and if needs be confront such behaviour. Such rules-based orders also give lawyers undue influence over policymakers and lead to "lawfare" between the political state, the administrative state, and the legal state, all of which are seeking power at the expense of the others. The result is a country in which parliament is nominally sovereign but in which its very power is contested by Whitehall administrators who gold-plate all and any such rules to increase their own power and constrain the political state, and lawyers who present what is often political as merely legal. The paralysis in British policy and strategy is thus hardly surprising.

Nowhere is that confusion more apparent than in British defence policy and the assumptions London makes about what defence Britain needs, and how much it will cost. Defence policy, and consequent defence strategy, also suffer from the kind of strategic and political virtue-signalling which is doing so much damage to Britain's ability to compete in a hyper-competitive world—competition that is ultimately existential. The consequence is that the word "strategy" has become the most over-used and mis-used word in the British political lexicon.

The consequence is a London that too often tries to give the impression of decisive action when in fact this action is far less than the desired appearance would suggest. Britain has thus become just another risk-averse European power, a kind of Belgium with nukes. The British government is in any case insufficiently "joined up" to generate the efficiency and effectiveness across government that contemporary security and defence demands. Rather, the great ministries of state—the Home Office, Foreign, Commonwealth and Development Office, and the Ministry of Defence—have become competing baronetcies of power, each with their own distinct political cultures, with the only shared belief an aversion to risk and the appeasement of dangerous reality. Oversight bodies such as the Cabinet Office and the National Security Council simply lack the necessary political heft to impose unity of purpose and effort, leaving Treasury economists as the ultimate arbiters of strategy. A so-called Fusion Strategy has been created, and a seventh advisor to the prime minister on national security (National Security Advisor) has been appointed, the first serving military officer, General Gwyn Jenkins. However, given the way Whitehall works, neither of these steps are likely to prove game-changers. It will be particularly hard for the National Security Council to fulfil its primary mission of overseeing a new "strategic cycle" to deliver more and work more cohesively across the whole of government. Consequently, national, security and defence strategies, particularly since 2010, have only pretended to balance the ends, ways and means of British grand strategy.

Britain today is not in the worst of all strategic positions, but its position is uniquely uncomfortable in that London has responsibility without authority let alone flexibility. London is forced to act by the need to maintain a critical relationship with the United States upon which Britain relies for so much, but is incapable of any real influence over Washington. Consequently, London is forced routinely to respond inadequately to crises made elsewhere whilst exaggerating its ability to act. One ship to the Red Sea, one battlegroup to Estonia, half a battalion to Kosovo, all in the hope that the crises do not deteriorate because if they do Britain has few credible and rapidly deployable reserves to draw on. In other words, the size and capability of any British force is dictated not by the situation on the ground but by the inadequacy of policy, with potentially disastrous results for those deployed.

The policy consequences of being forced into a "strategy" for which it simply does not have the means are profound to say the least. First, a retreat into a defensive *and* competitive groupthink in Downing Street, allied to a propensity to moralise to mask

Britain's strategic retreat. Second, the erosion of the vital link between British grand strategy, strategic objectives, and Britain's national instruments of power across the diplomatic, informational, military, and economic (DIME). Third, what passes for "strategy" is a metaphor for under-investment. Fourth, the progressive infantilisation by the elite of a mistrusted people, which not only destroys the trust between leaders and led, but also prevents London establishing an essential partnership with its citizens vital to keeping Britain safe.

The result is a Britain that oscillates between nostalgia and utopia, "led" by weak leaders who reflect that oscillation. A part of the problem is the fragility of government after years of under-investment, with decision-making in London seriously hindered by the influence of special interest groups and quangos (quasi autonomous non-government organisations). Britain has also changed radically. Once one of the most stable societies in the world, built on a class system that sacrificed social mobility for stability, Britain today is now a very mobile society, albeit insecure, fractured, divided and increasingly unstable. These divisions came to the fore during the several crises that have had a profound impact since 2008, most notably Brexit and COVID.

The profound divisions that exist within contemporary Britain could become unbridgeable. The political Right yearns for a past Britain that never existed, whilst the political Left dreams of a universalist society that will never be realised. If Britain is to compete, prosper and even survive in the twenty-first century there simply can be no place for either nostalgia or utopia. If strategic pragmatism is to be restored to the heart of Britain's way of doing things in the world, it must be a strategy that also reflects a new Britain. A Britain that collectively understands that in a dangerous and networked world all-of-state approaches are vital. A Britain that is sufficiently robust to cope with the consequences of shock. A Britain that understands and embraces power and purpose through the planning and efficient execution of a coherent national strategy which if properly conducted can mitigate the worst effects of shock by building resilience and redundancy into national systems and society. Above all, London must stop believing its own propaganda.

An essential argument of this book is that Britain has too often subsumed its vital interests in pursuit of vaguely defined values. The opposite is also at times true. There is a profoundly misplaced and often naïve belief in government about the benefits of "just-in-time" globalisation. Over many years London's naïve globalism and the false assumptions about power and policy it generates, not to mention the dangerous misperception that others share the same values, has undermined the competence and expertise for which the British state was once celebrated. This retreat from competence has not only exaggerated Britain's vulnerabilities and weaknesses, but it has also helped magnify and exaggerate the strength of autocratic adversaries, such as China and Russia. This tendency is at its most revealing in the attitude of successive governments to the City of London. The need to preserve at all costs the Golden Goose and its Golden Eggs has also led to the effective appeasement of those providing the eggs, be it the Chinese sovereign fund, Russian oligarchs, or money from a host of unsavoury sources.

Big Lessons from Ukraine

An interests-led British grand strategy would be precisely that. Take Ukraine. It *is* in the British interest for Ukraine to survive as a political entity, but not to go to war with Russia over Russian-speaking Donbas and Crimea. In 1938, as part of the settlement of the "Czechoslovak problem" Chamberlain negotiated away 20% of the then-Czechoslovakia as part of the Munich Agreement. Many in London accepted that war with Nazi

Germany was a very real possibility, but it was not in Britain's interest to fight such a war before British rearmament was more advanced together with its evolving military strategy of "steel before flesh" that would come to define the Western way of war during World War Two. The real betrayal of Munich was that it offered the Czechs, the Poles, and others no meaningful security guarantees. A similar deal over Ukraine would again be in the British interest, albeit with meaningful security guarantees from NATO, including future membership of the Alliance. Ukraine would lose 20% of its territory to the Russians, who would get to keep much of the Donbas and Crimea, which it has taken illegally by force, as well as Mariupol, a major Ukrainian grain port on the Black Sea.

This is partly because Britain and its allies have already given 90% of what they are able to give Ukraine, whether it is delivered as promised or not. It is also because the allies are never going to give the Ukrainians the equipment and operational freedom they need to defeat the Russians in a war in which what "defeating" the Russians amounts to is so unclear. The twenty-eight Western-trained and equipped Ukrainian brigades that formed the core of the 2023 summer offensive always lacked the military weight to break through the Russian defensive lines in the south and east of Ukraine. The Russian General Staff learned some painful lessons in 2022 and 2023, and the West took so long to deliver the relatively limited supplies of arms it had promised that the Russians had time to learn how better to fight on territory they controlled. Artillery has been the defining feature of this very Russian war and in early 2024 the Russian Army has five times the amount of 155-millimetre ammunition Ukraine possess. The EU also promised but failed to deliver 1 million artillery shells, mainly due to an inability to upscale rapidly European arms production, whilst Russia received over 1 million artillery shells between August and November 2023 from North

Korea, clearly with Chinese backing. In other words, Russia is winning the artillery war.

If keeping Ukraine alive is a vital British interest, but restoring Ukraine's 2014 borders, let alone its 1991 borders, is not, then what exactly should the strategy be? The aim should be to end the war as quickly as possible but enable Kyiv to negotiate eventually from a position of relative strength. Experience suggests trying to negotiate with Russians from a position of weakness is a fool's errand. That means keeping Ukraine in the fight in 2024, enabling Europe to upscale ammunition production so that in 2025 Russia is left in no doubt of the West's commitment to an equitable peace in Ukraine. However, such a commitment should mark the high point of Britain's ambitions.

There are several other factors that drive such a strategy. The most salient of these is the lack of a coherent Western grand war strategy since the February 2022 Russian invasion. The West has got into a habit of promising Kyiv "whatever it takes" whilst only giving Ukraine just enough weapons to prevent Russia from conquering the whole country. It is now clear few in the West believe in Ukraine's war aim of recapturing all the territory the Russians have taken. Frankly, the West has been self-deterred by the risk of a wider war with the Russians over the Donbas and Crimea, which in any case several European countries see Ukraine as having only borrowed from Russia. Worse, if Trump wins the November 2024 presidential elections Ukraine for the Americans might once again become a large distant country about which they know little. Hard but true.

How would the West justify such a retreat from the 2022 rhetoric (and it would be a retreat)? First, should there be negotiations (there are already extensive contacts with the Russians), the West would play up the "victory" of a rump Ukraine and the strategic failure of Russia to realise its original war aims. The West would also point to the reinvigoration of NATO. Second,

Berlin and Washington, and no doubt Brussels and Paris (not to mention in time London) would also say that by simply surviving as an independent country the sacrifice of so many brave Ukrainians was worth it. Third, they would hint how much cheaper it would be for the West, with Europeans to the fore, to rebuild Ukraine if they do not have to pay for the war-torn Donbas and occupied Crimea. Fourth, they would have secured an end to the killing by sacrificing some Ukrainian territory in support of Ukrainian sovereignty.

On the Russian side, Putin would no doubt present his vision of a "Novorossiya" and the rebuilding of a Russian Empire as vindicated and he would doubtless believe he had successfully completed phases one and two of its rebuilding. Phase one was the seizing of Crimea. Phase two, the successful if costly occupation of Donbas and Mariupol. Phase three? After he had rebuilt Russian forces, say by 2030, he would move to seize Odessa and cut Ukraine off from the Black Sea and seize all of Ukraine east of the river Dnieper, including the breadbasket. However, by then NATO would be ready for him, making it far harder for Moscow to contemplate attacking the Baltic States, the Nordic States, and/or the Arctic.

Clearly, Britain must better understand the longer-term geopolitical consequences of any fix imposed on Ukraine for short-term political and economic relief. If London and its allies fail to learn the real lesson from the Russo-Ukraine War—that Putin really is a militarist and an adventurer—and if Europeans again fail to properly rearm then all Europe would have gained through such a peace is a strategic pause. Britain should also join the EU in developing Ukraine's economy using perhaps the Korean Peninsula as a model. NATO would also need to offer what is left of Ukraine rapid membership of the Alliance the moment any such agreement comes into force. That begs a further question: would all NATO members sign up to Ukrainian membership?

Regardless, from the outset, NATO would need once again to become the credible warfighting force required to deter any further Russian military ambitions. As America focuses on the West's collective behalf on China, whilst nevertheless remaining a key member of the Alliance, Britain, France, and Germany must take up the slack in NATO.

As for British "strategy," the Russo-Ukraine War has been another example of London's almost romantic confusion of interests with values. In February 2022 Russia offended Britain's values to trigger London into a leading role in an anti-Russian coalition but because Russia stopped short of directly threatening British interests London, like most European capitals, set limits to its support for Ukraine. The West now has little option but to play a long strategic game in the hope that one day the four lost oblasts and Crimea might return to Ukraine, however unlikely. It will require strategic patience, but it is precisely a lack of strategic patience that so damaged all the recent campaigns in which Britain has been engaged. London also needs to be clear about its long-term grand strategic goals as part of the wider West, which include slowly bringing Russia back into the European *polis* and steadily prising it away from China rather than driving them together.

Other lessons from Ukraine? Britain and its allies are making the same mistake as previous campaigns this century, with no clear linkage apparent between the ends London desires or the ways and means to realise them. What is clear is that Ukraine will not "win" this war, even if "victory" was a defined end-state to achieve. They are a bit like Confederate forces at the Battle of Gettysburg in July 1863: unless something very profound changes Ukraine's heroic forces may be at or passed their high-water mark.

London has become overly hopeful that Ukraine will secure some ill-defined victory while ignoring Russia and her proven ability to absorb huge punishment and learn lessons before striking back. No plausible amount of Western equipment and support will now be sufficient. Worse, a Trump presidency would reinforce this reality. Allusions to the Battle of Britain in 1940 are utterly misplaced. In 1940, Britain was an island with the world's most advanced air defence system, one of the finest defensive fighters ever built, with an aircraft industry that by June 1940 was outproducing the Germans. Ukraine has none of these advantages.

As the US military are prone to say, "don't forget the enemy has a vote!" Many military "experts" who should have known better have focused on the tactical level, which President Zelensky and a succession of British leaders have taken at face value. Zelensky now has a huge problem because, like the Good Old Duke of York, having marched the Ukrainian people and its brave army to the top of the hill, it is now very hard for him to march them back down again. That is perhaps why Zelensky is making ever "bolder" claims about Ukrainian war aims, such as claiming much of western Russia as Ukrainian territory. If Britain and the West really were determined to help Ukraine expel Russian forces and return to even the pre-2014 border, it would mean giving the Ukrainians the necessary operational freedom as well as military capability to do so, but there is no evidence whatsoever that Britain and its allies are willing to do this.

From the British strategic perspective Russia has already been strategically defeated, a message that needs to be constantly hammered home. Moscow failed miserably to take Kyiv and has hardly expanded on what Russia already controlled in 2014. They have also lost hundreds of thousands killed and wounded in action, and because of Putin's actions NATO is bigger and more united than for many years and at last beginning to re-grow its collective military capability and all-important political will.

What next? Strategic logic would suggest that both Ukraine and the Alliance need to switch from the offensive to strategic defence because Ukrainian forces have reached what Clausewitz would recognise as their culminating point. This is not only because the critical supply of 155 mm artillery shells to the Ukrainians has faltered, but because their armed forces are exhausted, even if the West could give them the kit they seek. Any such defence would need to be guaranteed (unlike the 1994 Budapest Memorandum). Britain and its allies would thus need to give Ukraine the immediate security guarantees they require, and the equipment needed to defend the territory they now hold, as a prelude to NATO membership. Concurrently, the EU needs to focus on re-building Ukraine's economy so that in time she becomes a thriving well-defended EU state, not dissimilar to South Korea. Such action will also deter Russia from attacking any NATO state and give the Alliance time to reinforce its defences before Russia can reconstitute the forces it has lost in Ukraine. The West must not waste this opportunity.

The Book

The uniqueness of this book is not simply that it is a collaboration between a former Chief of the British Defence Staff, Britain's most senior military officer, and an internationally renowned strategic analyst, author, and Oxford historian. The book has also been supported by former senior figures, such as Lord Robertson and Tony Blair, who responded to an elite questionnaire, all of whom (thankfully) validated the thesis of this book: far from punching above its relative economic, diplomatic, and military weight, to use a time-worn phrase, Britain punches decidedly beneath it.

The book has two parts: "The Retreat from Strategy" and "The Return to Strategy." Each chapter focuses on what British grand, national, and defence strategy should address rather than what they do address, whilst using the gap between the two as the essential dramatic tension of the book. The book concludes with a vision of how *this* Britain might return to sound strategy, in all its manifold forms, but focused on a vision of a future British military force that is both affordable and sufficiently capable given Britain's strategic interests. Each chapter is thus a mix of analysis, experience, and insights at the strategic, tactical, and operational levels. The book opens and closes with two scenarios. The first envisages a British military defeat in 2031, the consequence of failed strategic thinking and defence planning and the continuation of the serial erosion of ends, ways and means which results in a catastrophically unexpected, but entirely predictable, defeat. The second scenario takes the British future force of the final chapter and affords the reader a vision of how an affordable and credible defence can be mounted if such power is seen again by London as a value rather than a cost.

Part One: The Retreat from Strategy

Chapter One, "Britain, Strategy and History," considers why Britain is retreating from strategy. David Richards' experience as Chief of the Defence Staff gave him first-hand experience of London's confusion of strategy with tactics, and values with interests, and how a culture of muddling through and managing decline have reduced Britain's strategic ambition. The chapter also considers the challenge of crafting effective strategy and compares Britain with its principal allies and partners, America, France, and Germany. It concludes by drawing some lessons for British strategy from the Russo-Ukraine War.

Chapter Two, "A Force for Good?," looks at the recent strategic failures in Libya, Syria, and above all Afghanistan, where David Richards was the only non-American commander of the pan-Afghanistan NATO International Security Assistance Force. The chapter measures recent British actions against four tests Tony Blair established in his 1999 Chicago Speech. The lesson is that without good campaign design and committed political ownership there is only so much a commander can achieve on the ground.

Chapter Three, "Smoke and Errors," considers the political culture in London that has put Britain's defences in such a perilous and impecunious position. As both Chief of the General Staff and Chief of the Defence Staff David Richards saw how little influence "Defence" had over London's political and bureaucratic firmament. The chapter also charts Britain's defence decline since the 2008 banking crisis, when wishful thinking in London led to defence being seen as little more than discretionary expenditure. Finally, it describes how such appeasement of a dangerous reality was masked by successive defence reviews that used integration of effort as a metaphor for unwise cuts.

Chapter Four, "Ends, Ways, and Has-Beens?," reflects on David Richards' experience as a force commander to consider contemporary consequences of London's strategic illiteracy on the forces he had the honour to command. It also sets current defence policy against three levels of that most cruel of arbiters—readiness—by charting the retreat from defence reality.

Part Two: The Return to Strategy

Chapter Five, "A Return to Strategy," considers how and why London should relearn the art of strategy to restore the vital link between the increasingly dangerous reality Britain faces and the ends, ways and means it must bring to bear to maintain peace and restore prosperity. Critically, it calls on London to use security and defence strategies for their intended purpose: to confront threat today and tomorrow, and not simply recognise only as much threat as the Treasury thinks Britain can afford now.

Chapter Six, "The Utility of (British) Force," considers what the British armed forces are and will be for in the middle decades of the twenty-first century. To that end, the chapter establishes NATO as the only viable framework for British defence strategy and calls upon London to be at the forefront of the development of a European-led Allied Heavy Mobile Force.

Chapter Seven, "Belgium with Nukes?," is the focal point of the book and faces up to the most important defence reality: on the current defence budget Britain can either be a credible nuclear power, or a capable conventional power, but not both. It also provides a vision of a British Future Force that would finally enable London to square this seemingly impossible defence circle.

Chapter Eight, "Tommy," looks to the future with hope. That which we are, we are... but not that which we can be. The book ends with a scenario of victory, "Britain Defended," in which the power of Britain's conventional forces, leading at the heart of a reinvigorated NATO in Europe, stands firm and ensures that the nuclear option remains precisely that!

PART ONE

THE RETREAT FROM STRATEGY

In the end, more than freedom, they wanted security. They wanted a comfortable life, and they lost it all—security, comfort, and freedom. When the Athenians finally wanted not to give to society but for society to give to them, when the freedom they wished for most was freedom from responsibility, then Athens ceased to be free and was never free again.

Edward Gibbon, *The History of the Decline and Fall of the Roman Empire*

1

BRITAIN, STRATEGY AND HISTORY

The development of a strategy (which should be underpinned by thinking strategically) allows the integration of different agencies of state. More importantly, it should cause a competition of ideas, which can be tested. The best ideas can then be incorporated into whatever strategy emerges. An important part of a strategy is that it provides transparency (to a degree) and buy-in for citizens, and it is an important way to telegraph support to allies and resolve to adversaries in the war of ideas we are involved in.

Major-General (Ret.) Mick Ryan in response
to the book's questionnaire

Muddling Through

Where to begin? Former Secretary of State for Defence and NATO Secretary General Lord Robertson answered that question succinctly:

> Launched with a fanfare..., [the Integrated Review is] now in the pigeonholes of the Cabinet Office, invisible to the public, and nothing seems to be being done to build a national consensus without

> which it has no chance of being a success. It should have sparked a necessary national conversation about where the country wanted to go, what the people needed it to do and to establish what we would do less of, in order to do more. If the Integrated Review is a strategy for the long-term future it cannot be the property of we transitory politicians of today.[1]

In February 2024, *HMS Vanguard* test-fired a Trident 2 missile off the coast of Florida in the presence of the Secretary of State for Defence and the First Sea Lord. It failed, at a cost of some £17 million to the British taxpayer. The last such test by a British nuclear-powered ballistic missile submarine, *HMS Vengeance*, also failed. Either the Americans are supplying "dud" missiles to the British or budgetary pressures on the Defence Nuclear Enterprise are beginning to show. What is increasingly clear is that on the current defence budget Britain can either be a powerful conventional military actor or a credible nuclear power, but it cannot be both.

How was Britain reduced to a position in which neither its conventional forces nor its nuclear forces are any longer credible in the core roles for which they exist—to defend Britain and its critical interests? Britain remains in global terms a very significant mid-sized power anchored in and to Europe. But the nature of the debate also reveals the extent to which British leaders themselves have little or no idea about what kind of power Britain should aspire to be. Since the 1956 Suez fiasco London has outsourced Britain's grand strategy either to the Americans or to Brussels, where at least the British were one of three major EU powers, along with the French and Germans, even if Berlin and Paris routinely refused to recognise Britain as such.

Now that Britain has left the EU, and the United States is amid profound domestic political turmoil, the British once again face a turning point in their history. For the first time in a generation London needs to think and act grand-strategically. Is

London up to the challenge? At the very least, London will need to relearn both the art and science of grand strategy, because in democracies any such strategy is inevitably a struggle between values and interests—a balancing act with which autocracies such as China and Russia need not bother given the indivisibility of interests and values in an untrammelled leader. In democracies, as in war, the nature of strategy is enduring, but its character is not. The quintessence of grand strategy in London is muddling along.

"Déjà vu all over again?" Britain today resembles pre-imperial Britain of 1707 and the Act of Union between England and Scotland, more than the truly Global Britain of 1907. Britain is a power of very significant regional-strategic significance, but is clearly no longer a global power, such as the contemporary United States. Britain's renewed grand-strategic modesty, after a 300-year imperial detour, begs a fundamental question: does contemporary Britain even need grand strategy, and if it does, what kind?

In fact, Britain has never needed grand strategy more. Strategy matters more to the relatively weak than the supremely powerful for which power alone can assure ends. British grand strategy, by contrast, involves and requires the generation, organisation, and application of still globally significant, but limited (and in terms of relative power ever more limited), national power in pursuit of what matters most to contemporary Britain: relative wealth and prosperity reinforced by relative security and stability. There are no absolutes in either power or strategy, nor is there room for reactionary nostalgia or progressive utopianism. If a clear, legitimate, and pragmatic direction of grand strategic travel is to be established, it must be supported by appropriate mechanisms of government and governance and the means available to realise the desired ends. In the past, grand strategy reflected the self-supposed grandeur of the country. British grand strategy was a settlement between a gilded aristocratic elite born into privilege and educated at the same schools and universities and a mercantilist elite focused on the accumulation of immense wealth. The crafting of grand strategy in the Britain of today is no easy task given the complexity of contemporary British society, the fracturing of power from which the British state suffers, and the hangover in London from that past. L.P. Hartley's famous dictum, "the past is another country, they do things differently there," is even more relevant today than when he wrote *The Go Between* in 1953.

The assumptions implicit in British grand strategy can be thus summarised. First, the United States will always be there to offset Britain's security and defence deficiencies even if the Americans are increasingly overstretched the world over and domestically distracted. Second, that Washington will continue to listen to a London that often talks the talk of power and strategy, but rarely walks the walk. Third, that despite Brexit, France and Germany will sooner rather than later welcome the British back into a "*trirectoire*" of power on the continent, but only if Britain pretends to have left the EU. Fourth, the English-speaking powers and the Commonwealth can act as an alternative source of post-EU power for Britain.

Unfortunately, the very idea of "strategy" (let alone of the grand variety) has become debased in London. All the recent strategy documents are as much glossy political sales brochures as policy-planning drivers. They offer cut-price, "special deal" security to an often bemused and increasingly cynical public and false power to increasingly dismissive allies and partners. Their "strategies du jour" reflect a political moment rather than enduring grand strategic fundamentals, designed to cover London's retreat into a political laager.

The signs and reasons are all too obvious if one looks closely enough. Repeated crises that are only partially dealt with; the masking of political failures with Pollyanna-ish expressions of optimism; the breakdown of trust between London and important sections of the population evident in the 2014 Scottish independence referendum, the 2016 Brexit vote, and the 2019 General Election; the obsession with diversity at the expense of unity and competence in government; the inevitable shift to the Left that is a consequence of diversity action; and the endemic short-termism to which successive British governments have succumbed. There are also more mundane reasons, such as that those involved in the drafting of strategy in London have an at-best partial picture and without strong leadership from the top seek to impose their own worldview on the process, and the fact that political appearance is far more important than strategic substance in London. The latter is often justified privately on the grounds that if British citizens really knew the state of national affairs the stability of society would be threatened.

There is also a patent lack of political consensus between the political state and the bureaucratic state, and a lack of strategic self-belief at the heights and heart of government. The iron grip of Treasury economists over government contributes to a culture in London which only recognises as much threat as government economists believe Britain can afford, and only in the short-term. The longer-term? Somebody else's problem. This is the very antithesis of sound strategy-crafting. Consequently, the very mechanism of strategy-making has become counterintuitive, with national security strategies normally followed by comprehensive spending reviews that often abandon the commitments made in the strategies in favour of short-term and incomplete prescriptions on the grounds of affordability. This causes the essential tensions at the core of British strategy: the need to do more in a world which is becoming increasingly fractious and dangerous, an ageing population that is increasingly a beneficiary of the public good rather than a contributor, repeated shocks to government assumptions, and an economy in which growth has essentially flat-lined compared with many others beyond Europe since 2008.

The Victorian Tyranny

Britain is still trapped by the tyrannical myth of Victorian greatness, a fleeting measure against which contemporary Britain always judges itself. And yet, even at the height of its Victorian supremacy, Britain was never a hegemon and London was never as strong as the impression it liked to give. The Victorian British Empire was always a ramshackle affair, a confluence of the industrial revolution in Britain with free trade at an opportune moment. France had been laid asunder by Napoleon's defeat in 1815, Russia was mired in its own Tsarist backwardness, America had only just begun colonising itself (and with the 1823 Monroe Doctrine effectively secured Britain's grand strategic rear), Austria-Hungary was Austria-Hungary (an empire much like the Forth Railway Bridge in that the massive bulk of its structure was there not to carry trains but simply to keep itself standing), whilst Bismarckian and Wilhelmine Germany did not yet exist.

Between 1815 and 1890 Britain had pretty much a free geopolitical hand and the instrument to use it—the Royal Navy. Lord Palmerston's famous 1848 dictum is worth quoting at length because it reflected a culture of thinking grand-strategically, rather than a grand strategy per se:

> I hold with respect to alliances, that England is a Power sufficiently strong, sufficiently powerful, to steer her own course, and not to tie herself as an unnecessary appendage to the policy of any other government. I hold that the real policy of England—apart from questions which involve her own particular interests, political or commercial—is to be the champion of justice and right; pursuing that course with moderation and prudence, not becoming the Quixote of the world, but giving the weight of her moral sanction and support wherever she thinks that justice is, and wherever she thinks that wrong has been done ... I say that it is a narrow policy to suppose that this country or that is to be marked out as the eternal ally or

> the perpetual enemy of England. We have no eternal allies, and we have no perpetual enemies. Our interests are eternal and perpetual, and those interests it is our duty to follow... And if I might be allowed to express in one sentence the principle which I think ought to guide an English Minister, I would adopt the expression of Canning, and say that with every British Minister the interests of England ought to be the shibboleth of his policy.[2]

Britain's relative power between the 1820s and 1880s was such that it by itself created a culture at the heart of government in which thinking grand-strategically became a reflex. Government's strategic assumptions were built on its understanding of power and its utility, an essential part of any effective grand strategy. Since 1945, and even more so since the 1990s, London has abandoned such principles of power for the vagaries of values, even at one point creating a "values-based foreign policy" that broke the relationship between power and interests. If policy could be dictated by whatever issue of that day held political sway, then so could strategy. Consequently, power became a dirty word linked to theft and oppression. If power was dirty so must strategy be, so any last vestige of grand strategy was replaced with meaningless grand values. The problem is that there can be no values without power because it is power which shapes strategy and structure, which, in turn, apply influence.

Mastery

For much of its history England, and then Britain, had to master strategy from a position of relative weakness and did so successfully often in the face of powerful adversaries. From the Valois kings of France, Philip II of Spain, Louis XIV and Napoleon to the German threat posed by Kaiser Wilhelm II and Adolf Hitler, the British "genius" was to use geography allied to statecraft and coalition-building that reinforced London's enduring strategy of preventing a European hegemon. Britain was the balancer because of its seemingly innate ability to balance power, strategy, and structure. As Britain's power grew in the eighteenth and nineteenth centuries it became the go-to master of strategy and power, the ultimate status quo power, that shaped and shoved the international system in its own interests and created an architecture of power that endures to this day.

On 31 May 1889, Queen Victoria signed into law the Naval Defence Act, which formally adopted the so-called Two Power Standard. This committed the Royal Navy to maintain twice the strength of the next two most powerful navies combined. Even in this statement of grand strategic British power a seed of doubt was being planted for the future that in time grew to become a rotten oak. On the face of it the Two Power Standard was a statement of British Imperial might, but it was also a recognition that the French and Russian navies could if they combined make life exceedingly difficult for an over-stretched Royal Navy, by choosing when, where and how to apply pressure the world over.

Statecraft can best be defined as the skilful management of state affairs in pursuit of competitive advantage. Perhaps the most notable example of grand strategic statecraft came at the Congress of Vienna between September 1814 and June 1815. Britain had just fought a two-front systemic war with Napoleonic France and the nascent United States of America. Through skilfully using the power of its navy after the October 1805 victory at the Battle of Trafalgar, the quality of its regular army, and the effective application of national wealth and statecraft, Britain was able to see off both threats. First, with the defeat of the Americans in the war of 1812 British North America was secured. Then, in 1823 the Americans established the Monroe Doctrine, effectively blocking the ambitions of any European power in the American hemisphere beyond those already established. Unwittingly, the Americans opened the door to the post-

1815 British Empire by effectively protecting Britain's strategic rear. Second, the Congress of Vienna gave mercantilist Britain a free hand globally by keeping Europe divided and France weak.

Whilst historians have tended to focus on Austrian Foreign Minister Metternich and French Foreign Minister Talleyrand in Vienna, they have under-estimated the impact of British prestige, particularly in the wake of Wellington's victory at the Battle of Waterloo in June 1815. Foreign Minister Viscount Castlereagh and the Duke of Wellington reinforced British prestige in Vienna and enabled Britain to ensure its interests would be protected, with Britain a kind of offshore arbiter of disputes. It was Britain that invented the concept of the balance of power, enabling London to act as guarantor without the need to station large and ruinously expensive armies on the continent. Britain could stand aside from most internal European disputes and continue its imperial expansion. This was the result of Britain's leaders having the vision and will to act grand strategically, with statecraft doing the fine tuning. The Final Act of the Congress, signed on 9 June 1815, cemented British dominance of the world for a century, and Britain's dominance of Western Europe until the Franco-Prussian War of 1870–71.

Defiance...

In August 1919, London established the Ten-Year Rule which said that Britain should draft estimates for the Armed Forces based on the principle that the British Empire will not be engaged in a systemic war for at least ten years. London still imposes upon itself a virtual Ten-Year Rule even though the possibility of a major systemic war within the next ten years is becoming glaringly obvious. Britain also championed disarmament to offset the cost of World War One and sacrificed interests for values as London vainly tried to establish the first truly rules-based global order, in which the arbiter of state power would be the "court of public opinion."[3] It failed, first because it was nonsense, but also because of American indifference. The policy was finally ushered to destruction by the rise of Fascist Italy, Soviet Russia, Nazi Germany, and nationalist Japan. In late 1933 Britain finally revoked the Ten-Year Rule and in February 1934 began to rearm even as it struggled out of the Depression. War clouds were again on the horizon after Adolf Hitler came to power in January 1933. By September 1939 the British economy was in certain important respects better prepared for war than Germany's. How did Britain do it? Between 1934 and 1939 Britain mobilised strategy, technology and industry as London finally accepted that another war with Germany was possible.

With the creation of the Dowding System in 1936, allied to the invention and deployment of Chain Home radar, by 1940 Britain had the world's most advanced air defence system, something which the Luftwaffe discovered to its great cost during the Battle of Britain. The Royal Navy also remained the world's largest until 1943 (when it was finally eclipsed by the United States Navy), although the British Army had lost much of the fighting power that had seen it become the dominant land force of 1918.

Despite the creation of a mechanized force in the 1930s the British Army was the Cinderella force of the three-armed services. This was quite deliberate: London, like today, did not see Britain as a continental power, precisely because it did not want to fight an essentially continental war. It was the downside of being the balancer, because it enabled those in power in London to foster the false belief that even whilst a free Europe was critical to British interests someone else (the French) would do the bulk of the defending. For the Baldwin and Chamberlain governments of the 1930s, the French Army was essentially the British continental army. Much like today with NATO, Britain's core military-strategic assumption (and hope) was that France would provide the land power needed to deter, and if needs be, defeat the Wehrmacht. Unfortunately, the French leadership and French Army of the 1930s were rotten to the core, riven by rivalries and with much of its thinking still locked in the national trauma of Verdun. Only Charles de Gaulle was a true military visionary, but he was only a colonel at the time.

After a difficult beginning, Britain fought a sophisticated but ruinously expensive war between 1940 and 1945, at a huge cost to its economy and its power. Post-World War Two an exhausted Britain went into rapid decline, even if until 1965 it was nominally the world's second largest economy. London also became strategically emaciated by confused assumptions about the role Britain could play in the world as the sun very rapidly set on the British Empire. There were some in government who believed Britain could continue to shape the world by standing on mighty America's shoulders, even if Washington understandably would have none of it. Others in London saw their role as little more than managing Britain's inevitable decline. They hoped to find a sanctuary for a more modest Britain in NATO and the emerging European institutions, and it was the latter who eventually won the argument.

If World War One was the beginning of the end of Britain as a global power, World War Two marked the true end. The tragedy was that between Pax Britannica and Pax Americana there was a power vacuum which, due to American isolationism in the wake of the 1919 Treaty of Versailles, the declining British and French tried to fill with empty institutionalism in the form of the League of Nations. Fascist Italy, and then more ominously Nazi Germany, Tojo's Japan and finally Stalin's Soviet Union all tried to impose themselves as Leviathans on world politics. Lessons for today?

Having become deeply indebted to the US during World War One, Britain faded as a world power while resisting tyranny during World War Two, with the result being Britain today. London has never really escaped the habit or memory of power, even if it lacks the fact of it. Britain behaves as if it is still a dominant status quo power when it is in fact little more than the recipient of decisions made elsewhere. Britain has become a commentator on power, semi-detached from it, lacking it, but expert on it. Consequently, Britain's need for a distinctive grand strategy is greater than ever, even if those responsible are perhaps the least able to understand the point: Britain needs to help shape the choices of others, something Churchill understood to his core. If Britain is to shape the choices of powerful allies, such as the US, Germany and France and use institutions such as NATO to that end, London needs a Plan. Muddling through is simply the mismanagement of decline because it makes Britain a victim of events not a shaper of them. In other words, it is precisely because Britain today is a "weaker" power that it needs a grand strategy.

... and Decline

In 1962, former US Secretary of State Dean Acheson caused a political storm when he rightly observed that "Britain had lost an empire but not yet found a role." Acheson's comment came in the wake of a rapid period of decolonisation and the 1956 Suez fiasco. The comment resonates as much today as then. Having failed to become part of the Franco-German-led European "project," abandoning the EU in 2016, Britain is endeavouring to reinvent itself as "Global Britain" by emphasising its links with the United States and former imperial dominions, such as Australia, which are today very different states to the ones Britain did so much to create.

The defining moment in Britain's post-war decline, and perhaps the moment Britain stopped being an imperial power, was the 1956 Suez Crisis during which Britain and France, with the support of Israel, tried to seize back the Suez Canal from Egypt's nationalist President Gamal Abdel Nasser. It was the hero of D-Day, US President Dwight Eisenhower, who effectively pulled the plug on this last imperial adventure with devastating consequences for Britain. After Suez Britain effectively handed its grand strategy over to the United States and began the long and painful process of trying to mask its own diminution by ever more shrill appeals to the Special Relationship.

France chose another option. Whilst London decided it would never again cross America, France decided it would harness "Europe" to balance America, for better or ill. It turned from containing the Federal Republic of Germany to harnessing it through partnership. The Franco-German Axis might have been born in the European Coal and Steel Community of 1950, but it was consummated in the Suez fiasco. At the same time, Britain hung onto the trappings of power even as it retreated from the substance of power: strategic pretence became the method of British statecraft, most notably through the privileged access Washington eventually granted to American-made nuclear missiles.[4]

Britain's January 1973 entry into the European Economic Community (EEC) accelerated its retreat from grand strategy. The British political class effectively handed over power, sovereignty, and strategy to bureaucrats in Brussels (including their own) in the name of "Europe." It was the great victory of the bureaucratic state over the political state because whilst it bolstered the former it constrained and divided the latter. For the first time in over 200 years British grand strategy was made elsewhere whilst British "strategy" was reduced to little more than seeking the middle ground between American power and French and German ambitions, whilst trying to constrain Euro-federalist ambitions. London lost the art of making grand strategy, and to prove its European *bona fides* London developed the habit of "gold-plating" Brussels' directives. This both reinforced the power of the bureaucratic state over the political state and masked the rate and scope of British decline. "Europe" was not the reason Britain became a parochial power, but it provided the alibi for it.

Today, the United States, once the Great Arsenal of Democracy, is beginning to face the same global pressures that were exerted on Britain in the late-nineteenth century, not least by the Americans. When President Biden came to power in 2020 after the turbulent years of the Trump administration many in Europe hoped the transatlantic relationship would return to some form of "business as usual." What they meant by that was American largesse would continue to enable Britain and its European partners to essentially free ride on the US for their security and defence. Nowhere was that sentiment stronger than in Britain, where London had failed to look beyond the politics of Trump to see the structural challenges affecting American policy, both foreign and domestic. Rather, London used Trump as an alibi to avoid facing the hard security and defence choices they had to make in the hope their assumptions would be restored.

Upon taking office in 1979 Margaret Thatcher was shocked to discover the civil service saw its primary mission to be one of managing Britain's decline. Little seems to have changed. The 2001–2020 campaigns in Afghanistan and Iraq, replete with echoes of Britain's imperial past, were in many respects the final nail in the coffin of British grand strategy. Today, Britain exists in a grand strategic vacuum with little or no idea of its role in Europe or the wider world. The consequence is "faux strategies" in which grand declarations lack meaning or substance, in which the ends, ways and means of strategy are never balanced, whilst values are routinely presented as interests and Britain's residual status as a nuclear power used as a metaphor for grand strategic vestige.

The retreat of Britain into such strategic pretence and the profound lack of strategic clarity and foresight it reveals came to a head in 2010, when then-Defence Secretary Liam Fox faced a £35 billion shortfall in the defence budget. The Cameron administration announced swingeing cuts to the British armed forces, even though London was in the middle of a major campaign in—of all places—southern Afghanistan. Helmand Province was in many ways the centre of gravity of the entire American-led campaign to create an Afghanistan no longer a threat to itself or anyone else, and hubristically Britain had taken on the challenge. Strategic Defence and Security Review 2010 began a process by which London built ever more risk into defence by scrapping capability as the demand for capacity grew. At a stroke, it scrapped the entire fleet of RAF and Royal Navy Harriers, including the carrier fleet and nine new Nimrod MRA4 maritime patrol aircraft (MPA) vital to the protection of the submarines carrying the nuclear deterrent. Fox's folly was in fact the work of the then-Chancellor of the Exchequer, George Osborne, and became the leitmotif of the Treasury-led retreat into Britain recognising only as much threat as London could afford, whilst also sending its armed forces on dangerous missions only partially equipped and too often under-gunned.

As Britain's leaders steadily withdrew from defence reality they also withdrew from political ownership of their own decisions, with profound consequences for the young men and women they put in harm's way. In Afghanistan, other equally weak but more canny Europeans simply limited the risk to their under-equipped and under-sized forces by insisting on restrictive rules of engagement and refusing to rotate their forces through eastern and southern Afghanistan. Britain, on the other hand, piled in. In January 2006, in the wake of Britain's profound failure in southern Iraq, the Blair government decided to send 3,000 British troops to Helmand, even though the Joint UK Plan revealed how little ministers knew about the complex tribal, political, religious, and criminal dynamics. The aim, the then-Secretary of State for

Defence John Reid said, was to use the force to assist civilians over three years to promote economic development, establish rule of law and create educational opportunities in a region of Afghanistan close to Pakistan, where there was little or no state presence. Reid even went as a far as to say he did not expect a shot to be fired.[5] Under the mantra of winning hearts and minds the "strategy," such as it was, had more to do with wishful thinking and virtue-signalling than effective strategy. Britain as a force for good but without the power for good.

A Changing America and a Changing Britain

The only reason Britain could continue to harbour such hubris was London's belief that the Americans would always underwrite it, whatever they said behind firmly closed doors. Today, there is a new hard strategic truth London must face. Despite their ability to project military project power, the Americans will not be able to be everywhere all the time in strength and across the full spectrum of twenty-first century threat and conflict. Power is relative and for a state to exert such influence it would need to be uniquely strong in relation to all other possible peer competitors. There may have been a moment back in the early 2000s when some Americans thought the US could be a unipolar power and act as the Global Policeman (even if many Americans did not want to), but 9/11, Afghanistan and Iraq, COVID and the rise of China have quickly proved such pretention to be illusory, if not delusional.

The United States is also changing. Many Europeans tend to view America through the prism of "the Greatest Generation," which in tandem with Churchill's Britain and Stalin's Russia won World War Two. They forget the isolationist Vandenberg America of the 1930s and ignore the extent to which the US is again fast changing. There were two telling trends in the 2020 US presidential election that London should have heeded. First, the percentage of white voters fell from 70% in 2016 to 65% in 2020. Second, the sheer scale of voting revealed a far greater engagement of minorities in the electoral process. This is to be welcomed. Political legitimacy in liberal democracies needs the greatest possible number of citizens to engage in the democratic process. Analysts too often tend to see geopolitics in terms of power indicators, which are often stripped down to size of a respective state's economy and the relative power of its armed forces. However, the ability of a state to apply power also rests upon a range of other, often intangible domestic factors. Whilst the power of the ageing "baby boomer" vote was again apparent in the 2020 election, their future is behind them. Twenty years hence the US will be a different place wearing a different identity and political complexion, and possibly have a quite different worldview.

Britain today is a paradox to Americans. Britain should be a very capable ally able to help relieve US military overstretch. Contemporary America is not unlike late imperial Britain in the 1920s. On the face of it, 1920 saw British power and influence at its zenith. Britain emerged from World War One victorious and in 1920 still possessed by far the largest navy in the world, the true measure of global power at the time. However, Britain was also mired in debt, not unlike the US today which faces a budget deficit of some 16% GDP, the largest since 1945, and a national debt fast approaching $35 trillion, some $10 trillion greater than the 2024 US economy of $25.5 trillion, (this is some 120.37% of GDP having been 33.27% in 2000).[6] Britain's national debt had grown from 25.36% in 1913 to 183.27% by 1923, much of it owed to the Americans.[7]

For the Americans the message is clear: if the US still has the will and political cohesion to lead the free world it can do so, but only in concert with committed and capable allies. In the Indo-Pacific that will mean deeper ties with Australia, South Korea and

Japan—and possibly India. As for Europe, the Americans need a strong Britain in a strong NATO and an Alliance of capable allies willing and able to properly share risks, costs, and burdens. If not, then in time NATO will simply prove too costly for the Americans to maintain. However, if such a new NATO is to be realised, *this* America must want to lead and be willing to continue to bear the costs of such leadership, which will remain substantial. The alternative? Look at Britain. A century ago, London's writ ran the length and breadth of the world. Today, London's writ hardly seems to run the length and breadth of Britain.

Britain is also changing, but then again it always has. The shift in Britain away from imperialism and towards disarmament was not just a consequence of the sacrifice of World War One. With the seizure of power by the political leaders of the bourgeois and working classes a British world view began to emerge that was very different from that of the Patrician order of old, the implicit story in the television drama, *Downton Abbey*, which any fan would recognise. In the early 1920s Britain was as deeply divided politically as it is today. The 1918 Representation of the People Act and the 1928 Representation of the People (Equal Franchise) Act extended the franchise to all men and women over the age of twenty-one. With two strokes of the parliamentary pen the age of High Victorian Aristocratic Imperialism (of which Churchill was very much a part) was effectively ended. The political and strategic consequences were profound.

Consequently, Britain's retreat from Empire accelerated because of the changing nature of Britain itself, in what was perhaps the first great struggle between imperial globalists and social nationalists. The Great Depression then further accelerated change in the global, political, and social order, just like the politics and economics of COVID have today. The change showed itself most clearly over the question of Britain's role in the world, in particular what was then termed Indian Home

Rule. Gandhi, Nehru, and others were successful (eventually) in agitating for Indian independence, but what is not often remembered is the support for such independence in Britain itself.

Masked by Britain's subsequent role in World War Two, historians too often overlook the lack of appetite for empire in much of 1930s Britain, mired in poverty and class conflict. With the political empowerment of the working class, both men and women, British politics rapidly became focused on the domestic struggle between entitlement, capital, and labour, as it is again today. Such tensions took the form of the 1926 General Strike and the rise of the Trades Union Congress. In contemporary social-media-driven Britain, they are reflected in culture wars, entrenched politics of identity, and the demand for far greater political and real investment in promoting racial and social equality. There is also the huge task that London faces in modernising the very fabric of Britain after decades of neglect.

Masters of Hot Air?

In January 2023, Sky News reported that,

> A senior US general has privately told [then] Defence Secretary Ben Wallace the British Army is no longer regarded as a top-level fighting force, defence sources have revealed. They said this decline in war-fighting capability—following decades of cuts to save money—needed to be reversed faster than planned in the wake of Russia's war in Ukraine. The source went on, "Bottom line... is that the service is unable to protect the UK and our allies for a decade." The sources said Rishi Sunak risked failing in his role as "wartime prime minister" unless he took urgent action given the growing security threat posed by Vladimir Putin's Russia. This should include increasing the defence budget by at least £3 billion a year; halting a plan to shrink the size of the army; and easing peacetime procurement rules that obstruct the UK's ability to buy weapons and ammunition at speed.[8]

Such statements are emerging from America because even the mighty United States is now grappling with the inevitable trade-offs between ends, ways and means that strategy demands. This is as evident in American strategy documents as it is in their British equivalents. The primary mission of the 2022 National Defense Strategy (NDS 2022) was to shape and size the US future force and the budget that pays for it. To do that it linked resources to strategy and force and it was noticeable that for the first time in such documents it also placed a particular premium on the support the US needs from allies and partners, most notably through NATO. This marked shift in American strategic doctrine reflects the simple facts of contemporary and relative grand strategic power and the growing over-stretch from which US forces and resources suffer given the rise of China in the Indo-Pacific. Some of the language in NDS 2022 would be recognisable to those who drafted the British strategies: both documents imply a greater role for allies going forward in assisting both the US and Britain to meet their strategic goals and challenges, particularly in and around the European theatre, and to offset perceived deficiencies, some of which have become severe since the February 2022 beginning of the Russo-Ukraine War. The British strategies might mirror the American language but they lack the American substance.

It is also noteworthy that China and the Indo-Pacific were afforded a higher priority than Russia and Europe in NDS 2022, even though Russia is described as an "acute threat." The American defence strategy follows on from its 2018 forebear, which switched US strategic emphasis away from counterterrorism and back towards Great Power competition, which is why both the US Nuclear Posture Review, and the Missile Defense Review were incorporated.

There are also four US defence priorities noticeably absent from British thinking. Priority 1 concerns the pacing, sizing, and shaping of the US future force to meet the challenge of China. Priorities 2 and 3 both emphasise deterrence against "strategic attacks" and "aggression," and the need to "prevail (not win) in conflict when necessary," whilst Priority 4 calls for a resilient Joint Force. The American defence strategy also calls for a "defense eco-system," a complex network of civilian and military stakeholders and partners, not unsimilar to the British Shadow Factory Plan of the 1930s. Crucially—and Britain take note—increased resilience is not limited to deployed force protection, but also applies to the US home base with the vital need for the Americans to be able to recover from the disruption caused by both "kinetic and non-kinetic threats." The American defence strategy also references sub-strategic threats, such as North Korea, Iran and violent extremism, whilst "trans-boundary" threats, such as climate change and pandemics, are to be adapted to.

Both the American and British defence strategies also reveal dangerously misleading assumptions about each other. Whilst NDS 2022 reveals the extent to which the Americans will increasingly rely on capable allies to realise their strategic assumptions, the British assume the Americans will always be in Europe in strength. The British thus still believe they can afford to merely pretend to be militarily capable. Where the 2023 Defence Command Paper becomes particularly light by comparison with its American counterpart is the vision for a future force. The US future force is to be built on three principles that demonstrate a clear link between American grand strategy, national security strategy, defence strategy and the enduring utility and purpose of American force: "integrated deterrence" and credible combat power (including nuclear forces); effective campaigning in the grey zone; and "building enduring advantage" by exploiting new, emerging, and disruptive technologies. The British strategies do not share the same level of ambition.

Not surprisingly, the difference is money. The American commitment to future defence dwarfs that of the British, including proportionately to the relative size of the two economies. The US is thus also choosing to be a relatively more powerful military actor than Britain by spending 3.3% of GDP on defence in 2024 compared with Britain's 2.1% today and (possibly) 2.5% in 2030, even though both London and Washington agree on the nature and scale of the threats and the vital need for armed forces that can operate together in the most extreme circumstances. America is committed to paying for such a force; Britain merely aspires to it. London has made a lot of political capital out of meeting the NATO target of 2% GDP on defence each year (of which 20% must be spent on new equipment). However, for several years this was achieved only by fiddling the accounts. In fact, between 2014 and 2023 Britain saw the lowest real term increase in defence spending (6.8%) of any NATO member. As Professor Malcolm Chalmers of the Royal United Services Institute says, "We're resting on the laurels of 2 per cent."[9]

The British, together with their allies and partners, need to quickly grasp the struggle the Americans are having affording and manning their prescribed force, which raises a fundamental and legitimate concern about wider Western strategy. If Europeans are really committed to the defence of the rules-based order that much of the world seems to reject, they will need to increase their defence budgets very significantly indeed. Therefore, for the NATO allies, with the British to the fore, the grand strategic runes of the US National Defense Strategy 2022 should be clear: if the US security guarantee to Europe is to be credibly maintained going forward Europeans, are going to have to share the defence burdens far more equitably by providing at least 50% of NATO's minimum capability requirements by 2030, which is probably the least the Americans will expect. Indeed, that will doubtless be one of the core messages of National Defense Strategy 2025!

Zeitenwas?

Should the Americans become ever more over-stretched there are two other European countries which on paper should be Britain's strategic partners in the defence of Europe—Germany and France. Together, Berlin, London and Paris represent almost 70% of all European defence expenditure and almost 90% of defence technological research and development. And yet, even before Brexit the defence strategic relationship with both was at best patchy.

The Bundeswehr is in a worse state than the British armed forces. As the *Spectator* writes,

> In an internal memo from November last year [2023], leaked to *Der Spiegel*, Mais [the head of the Bundeswehr] said that across the board the army had only about 60 per cent of the equipment it needs, "from A for artillery pieces to Z for tent tarpaulin (*Zeltbahn*)." Across a spreadsheet, he listed nearly 2,000 crucial items missing from Germany's arsenal, from piping and fireproof gloves to rather pointedly, a new fleet of Leopard tanks. The shortage list, Mais dryly concluded, "makes clear the diversity and small-scale nature of the challenges."[10]

The Russo-Ukraine War has also revealed the extent to which all three countries diverge on how to deal with the threat posed by Russia. As one leading German figure commented to one of the authors in December 2023, "I can say that things don't look great on [German] Ukraine support. We are still not seeing Ukraine's fight as all that connected to us [Germans]. It is about rules, order, solidarity, but not about us."[11] A leading French figure made a similarly telling comment when he wrote to one of the authors,

> What is currently believed is that Russia is to be reviled. However, as former French President Nicolas Sarkozy put it, "Putin was wrong,

> what he did was serious and resulted in failure, but once we've said that we have to move on and find a way out. Russia is one of Europe's neighbours and will remain so."[12] Without compromise, nothing will be possible, and things may be getting out of hand at any moment as mentioned by Chancellor Scholtz, "the West does not want the conflict to develop into a war between NATO and Russia by escalating the proxy war."[13]

Both statements reveal a marked contrast to how London sees the war in Ukraine.

Germany has eclipsed France as the most important European power, which presents London with both an opportunity and a dilemma. The May 2023 German National Security Strategy (GNSS) marks a step change in German foreign, security and defence policy in that it is really a statement of German grand strategy, albeit with two profound questions left unanswered: to what extent does the strategy mark a strengthening of the German instrument of hard power in German policy, and to what extent will this *démarche* lead to enduring change in line with British interests?

The GNSS was designed to act as a benchmark for current and future administrations in Berlin as well as for Germany's allies and partners, including Britain. It marked an important shift in German policy driven by the Russian invasion of Ukraine, the consequent *Zeitenwende* speech by Chancellor Scholz, and the SPD, Green, FDP Coalition Agreement of November 2021.[14] Certainly, new German strategy better aligns the ends, ways and means of Berlin's security and policy to meet extant and emerging threats than previous such efforts. Equally, and as in Britain, the defence bureaucratic politics in Berlin are illuminating. Whilst the Auswärtiges Amt (Foreign Ministry) was in the chair the strategy was to have been little more than a traditional statement of Germany as a friend to all, with mercantilism the implicit core of Germany's strategic outreach. However, the

Russian invasion of Ukraine led initially to the collapse of German assumptions about security in Europe and Berlin's profound belief that European stability is only possible through co-operation with Russia. But Germany is stuck. To illustrate this, there is the sorry tale of Berlin's promise to permanently station a combat brigade of 5,000 soldiers in Lithuania. As of January 2024, only thirty soldiers had been sent whilst one of the two tank battalions promised will have no tanks until 2026 at the earliest.

The German defence strategy has three elements, all of which share parallels with its British counterpart: the establishment of broad goals and interests, a description of a broad range of threats, and with whom to deal and how. Indeed, it is noticeable the extent to which the GNSS is cast in the tradition of US and UK security policy, with hard power playing a noticeably more important role than hitherto. Berlin also implies that Germany would be willing to reinvest in its security relationship with the US, which is perhaps why France is pushing EU military strategic autonomy so hard (for which there is little support in Berlin). There is also a clear desire to "Trump-proof" German security and defence policy in case Donald Trump is re-elected in November 2024, something which the British equivalent does not address.

Furthermore, unlike London, where HM Treasury imposed an iron grip over the drafting of Britain's recent security and defence strategies, the German Finance Ministry is supportive of the planned defence boost. The threat analysis in the German strategy for the first time talks of an "aggressive Russia," and the "systemic challenge" of China, even if Berlin prefers to "de-risk" rather than "de-couple" from such challenges. It also places particular emphasis on engaging the "fragile global south" and dealing with climate change. There is also clear language on NATO as a nuclear alliance and Germany's responsibilities therein.

However, that was then, and this is now. Paradoxically, the failure of Moscow to destroy Ukraine has also seen Berlin backtrack on its shift to hard power, with Chancellor Scholz rowing back from his *Zeitenwende* commitments.

If Berlin's defence aspirations—for that is what they are—were ever followed through Germany would eclipse Britain as Europe's foremost military power for the first time since World War Two. Berlin calls for the Bundeswehr to be rehabilitated with Germany moving towards spending 2% GDP of its $4.2 trillion economy on defence; enhancing the readiness of its armed forces is particularly important to it. After all, readiness is a key arbiter of genuine military capability. But herein lies Berlin's dilemma in that the new strategy was first and foremost an exercise in bureaucratic politics, a German sheep in a German wolf's (albeit a tame wolf) clothing. Throughout the drafting process, which took place at a particular moment in European history, the BMVg (German Defence Ministry) temporarily had the upper hand over the Auswärtiges Amt because of the war in Ukraine.

Berlin is thus a giant standing on the banks of the Rubicon and the challenge will be whether it will be willing to wade across. On 14 February 2024, German Defence Minister Boris Pistorius announced that for the first time since the 1990s Germany would in financial year 2024–5 spend 2% of GDP on defence by allocating €71.4 billion ($78 billion) to defence. However, as in Britain, much of the increase will be of the Special Military Fund, which will rise from €100 to €300 billion, with no clear indication of on what or when the money will be spent. The fund is also being eroded by defence cost inflation and is in any case merely a down-payment on the rebuilding of a severely hollowed out force. On top of that the major party in the governing coalition, the SPD, remains profoundly split over the issue of defence and, more importantly perhaps, German relations with Russia. If the cultural change implicit in Berlin's

National Security Strategy survives German politics, it will be in no small measure due to a persistently aggressive Russia. It is also clear that for all the rhetoric there are still very marked limits to the level of support Germany is willing to give Ukraine despite Berlin eclipsing Britain as Europe's biggest benefactor of Ukraine, at least on paper.

And it is that, ultimately, which is Germany's biggest test: is it a real power or an "at least on paper" power? The practical military issues Germany must also overcome reflect those in Britain, with the immediate priorities being the improvement of fielding times of equipment to the Bundeswehr and replacement of munitions sent to Ukraine. Both issues will require long-term contracts being granted by Berlin to arms manufacturers both at home and beyond. The good news is that for the first time since West Germany joined the Alliance in 1955 Berlin has passed a new law that states that funds will be provided to ensure Germany fulfils NATO Planning Goals, something the British have tended to observe in the breach.

Plus ça change

France is the country that in principle at least should, along with Britain, retain something approaching a residual grand strategic culture. Both are former imperial powers, which fought together during two world wars, both are nuclear powers and spend comparable amounts on defence. They are also Permanent Members of the United Nations Security Council with roughly the same sized populations at a similar level of development facing very similar threats and challenges. London and Paris do indeed share many grand strategic prescriptions, but Brexit, Channel migrants and the 2021 Australia, UK, US (AUKUS) submarine deal also reveal the shallowness and eternally fragile nature of the Franco-British strategic partnership.

In 2021, the French Ministry of Defence produced a Strategic Update which established French security and defence priorities out to 2030. With the title, "Proven Degradation of the Strategic Environment," the analytical preamble to the Update was lucid—and for *les anglosaxons* ever so slightly provocative—and demonstrated the complexity and scale of the overlapping and accelerating geopolitical and geo-economic challenges with which France must contend. While the French nuclear deterrent will continue to render another "Great Power War" unlikely, for Paris the threatening overall picture requires major change in transatlantic security responsibilities and the scale of European defence efforts. In other words, Paris wants greater European strategic autonomy from the Americans with France to the fore. In addition to citing terrorism, growing Great Power rivalries, and mutating hybrid strategies in a panoply of threats and risks, the Strategic Update also highlighted new competitive domains like hybrid submarine warfare, with reference to the seabed sabotage of cables. The French believe such challenges deserve very serious consideration within NATO, but also bilateral dialogues, most notably with the only other European power with a strategic culture, Britain.

Whilst the Franco-British military-military relationship is relatively healthy, the political relationship remains at best fractious, and it is hard to see how the two can be harmonised. Paris envisages a leading French role in managing a cooperative but complex process of strengthening NATO via the acquisition by Europeans of more and better hard military capabilities. At the same time, Paris would like to increase European "strategic autonomy" from the United States through the EU. Such strategic acrobatics between NATO and the EU has remained a hallmark of French strategic thinking since the Schuman Declaration of May 1950 and has often left the British as exasperated as the French are with Britain. The Strategic Update did little to end latent suspicions in either London or Washington about French neo-Gaullism.

Given the continued centrality of US military power in the defence of Germany and much of the rest of Europe, not for the first time Paris finds itself swimming against the prevailing European current in the hope it can persuade its partners that the future is "European." But is it? If the American commitment to European security really is weakening then Paris may just be right, albeit in the wrong way. For all that, French security policy has also become increasingly pragmatic and recognises that NATO European states need to do more, and more coherently, and via a stronger, and more coherent European pillar within NATO. However, French functionalism is never far from the surface in French policy and strategy and whilst the British have always seen NATO as an end, the French still believe NATO is, in time, a means to an EU common (in fact collective, not common) defence. Experience with the Franco-British Combined Joint Expeditionary Force (CJEF) of which David Richards was the principal architect, continues to provide a promising vehicle to straddle the EU-NATO divide.

With Germany seemingly permanently unsure about its military role and with Berlin still facing a military-phobic electorate, France still harbours the hope that Paris can lead the way to a more ambitious EU Common Security and Defence Policy (CSDP) built on Permanent Structured Co-operation or PESCO. France remains determined to preserve its extensive defence, technological and industrial base but until the Germans sign up to such a vision, if they ever do, then the Americans will remain the go-to ally for most Europeans including France. The difficulties evident in the Franco-German-led future combat air system (SCAF to use its French acronym) suggest the Germans are far from comfortable with exposing their taxpayers to subsidising the French defence industry. France is thus hedging. On the one hand, Paris continues to insist upon autonomous European strategic capabilities so that Europe might decide and act autonomously from the US, whilst on the other Paris wants to maintain France's capacity to act independently and/or alongside the US. By re-establishing European Political Co-operation (or EPC), in which European states co-operate collectively on foreign and security policy outside of the EU, Paris has also moved to bring the British back into the European fold, even though France continues to find Britain a complicated military partner.

The Retreat from Realism

Applied grand strategy is ultimately about directing the best possible ways, and immense means, in pursuit of geopolitical ends deemed vital to the country via the use of diplomatic, informational, military, and economic sub-strategies that are as joined-up as possible. Any such approach begs a series of questions. What are Britain's grand strategic objectives? How does Britain achieve these objectives? Is there sufficient unity of vision, will and effort to realise such objectives? Are there sufficient resources available? What planning traction is needed, and where, when, and how? Is there appropriate machinery to translate strategy into action?

Applied grand strategy requires a level of grounded ambition allied to clearly defined goals built on a clear understanding of critical national interests, allied to the pragmatic application of power through sound organisation and sustained investment. It is short-termism that kills grand strategy, and it is short-termism and the tyranny of political appearance that are endemic in the way British governments conceive of both strategy and policy. Worse, the wielding of influence and the pursuit of the national interest have become contentious issues in contemporary Britain, something other states find baffling. Therefore, the current debate over Britain's role in the world is not so much about Britain's role in the world, but rather the very nature of Britain itself.

It is a struggle taking place at the highest levels of the High Establishment, in which diversity is presented as strength rather than division.[15] The consequence is an increasingly unworldly worldview in government, which is the core of the contemporary struggle about how and where best to use Britain's still not inconsiderable power in the contemporary world. On the face of it, bundling security policy, defence policy, development policy and foreign policy into one concept of power appears to promote both greater efficiency and effectiveness of British statecraft, but in practice it masks the battle in government over the very nature of that statecraft. If national power is the amalgam of all capabilities and capacities a country can bring to bear, then it must first be conceived and organised at the highest level of government based first and foremost on the British interest.

London must also accept that Britain is not what it was but equally end the obsession with exaggerated decline. The 2008–10 economic and financial crisis finally ended any pretentions Britain might have had to be a world power. The banking crisis denuded British national wealth in ways not dissimilar to two world wars. Britain was effectively bankrupted and has been ever since. Worse, policies of austerity allied to attempts by London to mask the cost of political misjudgements also reinforced the tendency towards strategic pretence and the abandonment of national strategy. Contemporary British weakness is also the consequence of a vacuum of ideas and lack of will to confront realism, underpinned by a lack of strategic creativity and imagination in a mediocre political class which long ago handed grand strategy to the Americans and the EU. This retreat from realism is reinforced by a loss of direction evident in the defensiveness and muddle over the balance to be struck between Atlanticism and Europeanism, and soft and hard power. These profound tensions first became apparent in the Strategic Defence and Security Review 2010 (SDSR 2010) in which London tried to square an impossible circle, cutting defence by 7.7% in the middle of a major military campaign and shifting the cost of the nuclear deterrent from the Contingency Fund to the defence budget. This cut a further 6% per annum from the budget for conventional British forces. In other words, the Cameron government cut close to 14% from the British defence budget at a stroke. The British armed forces have yet to recover. Should we ask the banks to pay?

The 2008 banking near-collapse also saw the British economy shrink by some 7.8% between mid-2008 and the end of 2009, whilst the 2020–22 COVID crisis saw British public debt soar from around 80% GDP to over 100%.[16] In other words, London finds itself doing far more with far less. Given the nature and scale of emerging threats and challenges, unless the required ambition explicit in British strategy can somehow be afforded the ability of Britain to influence events will diminish further and rapidly. Closing this strategy-affordability gap will require a new approach to security and defence that will in turn demand much greater synergy across government and from top to bottom. That reality, at least, is implicit in the integrated reviews, with one notable lacuna: true synergy between the departments of state in Whitehall will only be realised if driven by an appropriately influential and weighty department of state, such as a truly powerful National Security Council.

Britain got itself into the current mess because it confused doing good with doing what is necessary, and what is vital with what is desirable. Therefore, to do good Britain must again learn to be tough enough to be tough and put the greatest effort into those areas most likely to secure its national interests, even if that also means saying no to the Americans from time to time. The great paradox of British security and defence policy is that by becoming so relatively weak compared with the Americans while making so much of British policy reliant upon them,

Britain has placed itself at the mercy of an increasingly capricious America. London *should* be close to Washington, but also decidedly and purposively strong enough to say no. By saying no, Britain will matter more to the Americans.

Too often British leaders have either sought refuge in America's sometimes dangerously one-dimensional strategic prescriptions, or deliberately neutered British influence by embracing decline. This absence of leadership is such that not just Britain's place in the world is open to question, but possibly the very future of the United Kingdom. Therefore, the need to re-establish some level of British strategic autonomy is not only pressing for Britain, but for Europe as well. That requires of London a much more systematic approach to the conceiving, crafting, and application of British power.

British grand strategy? The focus should be on the overwhelming totality of British political and military means for maintaining peace and stability on the European continent, taking pressure off the Americans world-wide, and leading that European effort with France and Germany. Then Britain would be applying affordable means in pursuit of clearly vital ends via a host of manageable ways. The 1938 Munich Agreement is a warning. In March 1939, Hitler broke the agreement and annexed the rest of Czechoslovakia which had been made defenceless by Munich. World War Two became inevitable. Failure to prevent war saw British defence expenditure rise from 3% of GDP in 1933 to 40% of GDP in 1940. When deterrence fails the costs are extremely high in both lives and geld. In other words, if Britain really wants to prevent war in Europe it must invest in the forces and resources to do so.

2

A FORCE FOR GOOD?

How the UK can best advance its interests in a world characterized by geopolitical volatility depends on the quality of its strategic thinking. This is more Darwin than Churchill. This is about survival of the fittest. It is not the strongest that comes out of the fight well, but the most adaptive.

Professor Rob de Wijk in response to the book's questionnaire

Something Must Be Done...

What happens when the ends, ways and means of Grand Strategy are poorly thought through, incompetently applied, and inadequately resourced? A review of Britain's failed campaigns reveals the consequences all too clearly. The tragedy of the chaotic August 2021 evacuation from Kabul Airport had its roots in Tony Blair's Chicago 1999 speech and the potential for strategic over-reach of liberal humanitarianism, if not carefully aligned with British interests, and carefully managed by at the highest levels of government. In April 1999, Tony Blair unveiled his "Doctrine of the International Community" as part of which Britain would act as a "force for good" in the world. In Afghanistan in 2001 the gap between aspiration, strategic application and tactical implementation became clear. Four years later in Iraq the gap had become so wide that the British doomed themselves to failure. It was the triumph of hope over experience. In the wake of the 9/11 al-Qaeda attacks on New York and Washington, anger and reflex trumped strategy and consideration and in the absence of cold analysis ushered in twenty years of British and wider Western failure, much like Israel's action in Gaza today. The result is the London of today which is so profoundly unsure of itself and its role in Europe and the wider world that it often retreats into hyperbole. Afghanistan and Iraq were the grandest of grand strategic confusions of values with interests. They were the ghastliest examples of inflated ends being imposed on insufficient means, forced to undertake ill-considered, value-driven, and often incompetent ways by a distant London driven by internal high-political and bureaucratic politics rather than the needs of the campaigns. London retreated into box-ticking, irrespective of whether the box was more relevant to people in Basra, Helmand, or London.

All of Britain's recent campaigns in Afghanistan, Iraq, and Libya, as well as its Syrian and Ukraine strategies, have suffered from the same profound tensions: affronted values leading to exaggerated demands for action but a rapid loss of strategic patience and political interest when the issue at hand is not seen to directly affect Britain's critical interests. This confusion between values and interests leads London to commit too early to ill-thought-out grand strategic ends, often made elsewhere, with operational and tactical means that Britain no longer possesses. Thereafter, a failure of strategic communication of any kind of meaningful plan leads to the steady erosion of support by the public for the campaign. This is exacerbated by the strategic demands of a campaign that requires a quasi-wartime footing, but the political need to maintain the armed forces on a peacetime footing. Such disconnects are further reinforced by defence planning assumptions which are wholly inadequate, the cutting of deployable forces during campaigns because of domestic economic and financial pressures, a consequent and creeping enforced reliance on the United States for support when campaigns become too big for Britain, and the abandonment of high-level political ownership in London as soon as the campaign inevitably becomes a political liability. Thereafter, responsibility for open-ended campaigns in which neither success nor an end is apparent is shifted inexorably and inevitably onto commanders and officials on the ground. At the same time, London cannot resist using its "long screwdriver" to insist upon political *démarches* that might be relevant in London but not in the field. Add to all the above the failure to invest in sufficient logistics and enablers whilst hollowing out the deployed force and the result is what is self-evident: British failure.

London also uses a series of "get out of jail cards" to mask failure. The first is to blame the Americans, who indeed must share a lot of the blame for recent Western failures, even though it was the Americans who filled many of the gaps in the British effort. The second is to constantly reframe the definition of success throughout a campaign to mask self-evident failure. The third is to measure input whilst studiously avoiding the measurement of outcomes. Finally, despite the constant refrain that "lessons will be learnt," the real lessons are not because they are too politically and financially expensive.

When David Richards became Commander, International Security Assistance Force, (COMISAF) in Afghanistan, faced with such tensions he repeatedly questioned American, NATO and British strategy and how they could be translated into effective tactics—but usually with little effect. There seemed to be little or no link between strategy-making in capitals, such as it existed, and the tactical reality on the ground. He and his successors devised their own linkage between strategic and tactical levels through operational level campaign plans that had little traction outside Afghanistan. Despite accepting the logic of Richards' arguments, politicians back in Washington, London and elsewhere simply refused to take high-level political ownership of the tactical implementation of the most important strategic stabilisation and reconstruction campaigns that the West had engaged in since the end of World War Two.

The strategic and tactical consequences of leaders in London distancing themselves from a campaign they had ordered were profound in that the ends, ways and means of campaign design and delivery were never properly synchronised. It also became painfully obvious to Richards that decades of mismanaging decline in London and trying to conceal it had left a huge void between what Britain needed to do and the ability of London to do it. In August 2021, the Afghanistan campaign reached its strategic denouement and a chaotic withdrawal at Kabul Airport ensued. Even then political leaders focused on, and at times seemed to revel in, a tactical withdrawal whilst ignoring the hard big strategic truth: Afghanistan was yet another complete failure for Britain and the wider West and would have profound geopolitical implications. One of these implications has been Russia's invasion of Ukraine. The picture of a West withdrawing in chaos—and even this only made possible with the cooperation of an "enemy" who had killed and maimed thousands of Allied soldiers and tens of thousands of innocent Afghan civilians—will resonate for many years across the world, most notably in the Global South.

Even as British forces began to come under pressure in Afghanistan Britain invaded Iraq as part of a US-led coalition. In 2003, David Richards was the Assistant Chief of the General Staff and an occasional member of the Chiefs of Staff Committee. He observed Western political leaders at close quarters. Both President

George W. Bush and Prime Minister Tony Blair had a relatively clear strategy for Iraq in 2003, but their tactics were (not for the first time) hopelessly flawed. There were also marked limits to Britain's influence. On one occasion, during a visit to Ambassador Paul Bremer, the US Head of the Coalition Provisional Authority in Baghdad, Richards tried to reverse the US decision to make the Ba'ath Party illegal, and stop the dismantling of the Iraqi Army and police. It became rapidly clear that despite the sacrifice and commitment of its armed forces Britain had little or no influence over the Americans. Napoleon may have said that one should never interrupt an enemy when he is making a mistake, but it was far harder to interrupt a friend. Britain, as always, backed down. The failure back in 2003 to properly forge sensible ends, and the best ways and means to seek them, led to a very long, bloody and tragically drawn-out process.

Where the Rubber Hits the Sand

David Richards was the only non-American commander of the post-2006 multi-national International Security Assistance Force (COMISAF). Already small by international standards, in 2010 the British Army was reduced by 7,000 personnel, whilst the Royal Navy and Royal Air Force were reduced by 5,000 personnel each during the campaign. At a stroke, the government of David Cameron effectively destroyed Britain's defence planning assumptions and broke the link between Britain's grand strategy and the size and utility of its armed forces. It profoundly weakened one of the critical means the British state needed to achieve national ends. The campaigns in Afghanistan and Iraq are thus case studies in Britain's retreat from strategy and London's withdrawal into a political fallacy in which defence has become little more than yet another discretionary investment in the public good. Britain also sent a message to the world: for the first time in centuries Britain would only recognise as much threat as its political leaders believed it could afford after other more politically compelling commitments such as health, education, debt, and deficit reduction. Whatever happened in the world beyond Britain, London was not for turning. The Americans were deeply unimpressed, whilst China and Russia took note. Britain, it seemed, no longer cared.

What a difference a decade made. On 7 May 2000, at the urgent request of the United Nations Secretary-General Kofi Annan, Prime Minister Tony Blair ordered British forces to intervene in a bloody civil war raging in the West African country of Sierra Leone. Operation Palliser was the largest joint operation conducted by British forces since the 1991 Gulf War. The political aim was twofold. First, to establish control of both the capital Freetown and Lungi Airport to enable the evacuation of British citizens and others in the face of an imminent threat. Second, to support the United Nations Mission in Sierra Leone (UNAMSIL). Ten days later, on 17 May, elements of the Revolutionary United Front (RUF) attacked British forces near Lungi Airport but were driven off with significant losses and their leader, Foday Sankoh, was captured. In September 2000, a ceasefire was in place and the RUF began to disarm as British diplomatic power exerted pressure on Liberia to end its use of conflict diamonds to bankroll the RUF. Following the ceasefire, a Disarmament, Demobilisation and Reintegration process began and in September 2001 the last British training teams were withdrawn and replaced by an international team with reintegration of former rebels almost complete. Operation Palliser was as clinical an exercise in the utility of British military power as London had dared hope. Then came 9/11.

Operation Palliser succeeded because it was a partnership between Sierra Leone's President Alhaji Kabbah and then Brigadier David Richards. It was an intelligent, well-planned, and textbook use of several British instruments of power as a force for good in pursuit of a clearly defined and achievable end at a manageable, if very real, level of risk. It was a mission that fitted perfectly into Tony Blair's famous April 1999 Chicago speech in which he established four tests for the use of British power. First, is the case for using British military power clear and have all diplomatic channels and options been exhausted? Second, can military operations be undertaken that are both prudent and have a reasonable chance of securing their political and humanitarian goals? Third, is Britain prepared for the longer-term consequences of such action? Fourth, are there clear national interests at stake?

Operation Palliser fulfilled all those criteria for action in that after the inevitable prevarications of the Whitehall system Blair finally deemed it important enough, doable enough, achievable enough, and that British forces were capable enough and sustainable enough to justify action—albeit with Brigadier Richards employing a creative interpretation of his mission.[1]

Contrast Operation Palliser with the far more ambitious and complicated campaigns in Afghanistan, Iraq, and Libya, all of which should have been seen from the beginning for what they were: over-reach. Since 2000, Britain has been involved in several major military campaigns in which few if any of Blair's criteria have been met. Worse, the means for achieving the stated ends via a variety of ways have been cut mid-campaign. Britain's interventions in Afghanistan, Iraq and Libya were always going to involve risk, but political decisions made by London destroyed the strategy on which the campaigns rested and thus doomed them to failure. London blamed the Americans, but the real causes of British failure were far closer to home.

Open Ends, Narrow Ways, Few Means

Iraq, Libya, Syria, and Afghanistan are now all testaments to twenty-first-century Western strategic incompetence for which

Britain must take its full share of responsibility. The consequences of repeated strategic failure are a loss of influence over both allies and adversaries which, taken together with Brexit, has seen a precipitous loss of British strategic influence, most notably in and over the institutions which have long been the levers of British influence. As former Defence Secretary Des Browne said,

> my belief is that having found a role in the intervening period, Brexit, together with a number of external political pressures, has left us somewhat uncertain of what our global role really is, what it should be, and what role our current capabilities and ambitions will allow us to sustain.[2]

Therefore, no consideration of future British grand strategy can be made without first considering the lessons from Iraq, Libya, Syria and of course Afghanistan. The one lesson that Britain's leaders should have learnt from all these campaigns is that there can be no "victory" unless peace has been properly planned for. All of Britain's twenty-first century campaigns have one other thing in common: the commitment of British forces and resources far beyond anything envisaged, and as the true demands of a campaign became clear a stubborn refusal to invest in what was clearly needed to succeed militarily. This was compounded by a refusal in London and elsewhere to consider the longer-term costs for Britain of restoring a broken country in the wake of intervention.

Things move quickly when a regime cracks. In Iraq in 2003, and Libya in 2011, short-lived celebrations at the ousting of Saddam and Gaddafi respectively quickly turned into chaos. London and Paris seemed to simply assume that "victory" in Libya would bring with it stability, even though experience in both Afghanistan and Iraq had taught them otherwise. The trigger for an intervention in Libya was possible genocide in Benghazi but those reports turned out to be exaggerations, and, in any case, where were the critical British interests? With British forces already heavily committed London should have avoided such a campaign. Then-Chief of the Defence Staff David Richards argued the case for negotiating a ceasefire with Gaddafi from a position of strength to avoid an unstable and uncertain outcome in a highly charged and still very tribal environment. This was clearly inimical to British values if not to British interests. Once again, from a liberal London's perspective the goal of removing a tyrant was politically too tempting for such mundanity and Richards' attempts to apply some realpolitik to the decision-making process fell on deaf ears.

Ukraine is but the latest campaign into which Britain has jumped without proper strategic consideration for ends, ways and means, campaign design, and the need to stay longer term to reconstruct a country as obliged by the UN Charter which London claims to champion. Britain has been particularly unclear about its aims, role, and political objectives in undertaking military action.

Iraq

Operation Telic in 2003 was a prime example of unrealistic ends and inadequate means applied via ill-considered ways based on a host of false assumptions. It was hoped that Britain's involvement in the US invasion of Iraq would be similar to the 1991 liberation of Kuwait: short, sharp, and relatively bloodless for the British. However, between 19 March 2003 and 22 May 2011, the campaign saw 46,000 British troops deployed at a cost of £9.5 billion, with 179 killed plus many more wounded, both physically and mentally. Saddam Hussein was removed from power, but Iraq remains to this day highly unstable and a breeding ground for sectarian violence across the Middle East and beyond.

False assumptions led to a critically flawed campaign design. First, after the toppling of Saddam in Baghdad the Americans,

British and others hoped that all they would have to undertake was peacekeeping as part of a transitional period. They found instead they were forced to become peacemakers. Second, whilst the British were only too happy that they were not in the so-called Sunni Triangle and facing the kind of opposition that American troops faced daily, Basra and the surrounding area they were responsible for was also a challenging environment, populated mainly by Shia Arabs and thus open to Iranian interference.

The essential British problem was that there were too few British soldiers to make peace in a very large and fractious area. After the so-called warfighting phase was over, there were at best 12,000 British troops. A further problem for both the Americans and the British was the lack of Europeans willing or able to reinforce them. This was partially because of Europe's perennial problem: whilst there are roughly some 1.5 million uniforms in free Europe, only some 10% or so can ever be deployed. Of that 150,000 or so, only between 40,000 and 50,000 could be used for the kind of peace-making required in Basra, let alone Fallujah. Given that such missions remain at least as likely as high-end warfare, this is another reason why the size of the British Army matters. It is wishful thinking to believe the military-strategic pendulum will not once more swing back.

Both the Americans and the British suffered from a peace-making deficit throughout the Iraq campaign, compounded by the campaign in Afghanistan. The Americans and British were thus forced to hand over less "unstable" regions of Iraq, both in the north and south, to Iraqis before the political situation had been stabilized or Iraqi forces properly trained and equipped, simply so the Americans could focus overwhelmingly on the contested Sunni Triangle around Baghdad. Few if any regions were ever sufficiently stable for either the Americans or British to concentrate their respective forces. The only way the British eventually got anywhere near to concentrating sufficient force was to abandon wider peace-making and retreat into Basra, defeating the very object of the campaign.

In 2003, the British Army "only" had 123,000 regular personnel, which was already deemed too small for the scale of the campaign. In 2025, the British Army will be some 50,000 troops smaller at 72,500 (if that). Even in 2003, the British would have ideally possessed a further deployable force of around 30,000 in order to rotate at a sensible tempo the 12,000 deployed in Iraq, while retaining an adequate reserve against fresh contingencies, as indeed Afghanistan was rapidly becoming. Britain did just about manage the first few rotations albeit at a very considerable stretch, but the degradation of the force through attrition was alarming. This reached its nadir when it became clear in 2005–06 that Britain could not meet its obligations in Iraq and Afghanistan simultaneously, leading to an ignominious withdrawal from Basra. Today, no such campaign could even be considered given how small the British Army has become. It is too small to meet Britain's NATO commitments and far too small to undertake all but the most permissive of expeditionary operations. That begs a fundamental question: what is the British Army for?

During the combat stage of operations in Iraq the UK deployed 46,000 troops of which 25,000 were combat troops, whilst logistics only allowed for a force of 9,000 deployed over three months at full combat strength. The hope was that civilian contractors would replenish military stores via a "just-in-time" strategy. It did not work, and the Defence Logistics Organisation came close to collapse. It is a problem that persists to this day. Given that Britain's hollowed-out force represents roughly 60% of Europe's advanced expeditionary warfare capability there is clearly a problem and not just for Britain.

Despite the acute challenges they faced, British forces made the best of the situation they found themselves in, which was no surprise as the British Army was for centuries an imperial police force and was experienced in advanced constabulary expeditions. The British Army also has a lot of experience of patrolling in non-permissive bordering on high-intensity environments. Thirty years patrolling the mean streets of Belfast, Derry/Londonderry and the "bandit country" of South Armagh between 1969 and 2003 reinforced a skill set.

Together with the issue of force density (that is, not enough soldiers to meet the operational need), a major problem in Iraq for the British was that the nature of such operations places a great strain on operational logistics over time and distance. London simply did not plan for such a reality. And yet, not only were the British in Iraq for a long time, but they were also in Afghanistan and Libya at the same time, and always with inadequate forces. Moreover, the British were over-confident, particularly in the immediate aftermath of the invasion, believing that they knew how to operate in Basra because of their experience in Northern Ireland. Peace-making in Iraq was far more demanding than peacekeeping in Northern Ireland.

Libya

In 2024, Libya remains little more than a failed state. Britain and France helped it to fail. It is also subject to increasing Russian and Turkish interference. Libya should matter to Britain. Tripoli is only some 294 km/184 miles from both the EU and NATO. Unfortunately, from the outset of the campaign London and Paris were unclear about their role and how to achieve their objectives—to help the Libyan people establish a durable and legitimate political settlement.

Experience from Afghanistan and Iraq had by 2011 suggested that all parties must be involved in political reconciliation early to prevent an insurgency gaining ground. This did not happen because London and Paris failed to understand the role played by the four key tribes that supported Gaddafi. In the wake of the military "victory" Britain and France failed to act to prevent reprisal killings and humanitarian suffering, and failed to help re-establish a seat of government or a clear political timetable for transition. They also failed to disarm and rehabilitate the various militias, and to help rebuild key state institutions such as the armed forces, essential services, and the judicial system. Nor did they get the backing of the international community to lead the support and assistance effort. A new UN Security Council resolution was needed early to legitimise support from key regional actors, the Arab League, the African Union and, of course, the European Union. It did not happen.

It should have been obvious from the outset that Operation Ellamy would be yet more hope over experience. There are five goals established by the United States Institute of Peace and the US Army Peacekeeping and Stability Operations Institute that any state contemplating a major peace-making campaign must consider: a safe and secure environment; rule of law; stable governance; sustainable economy; and social well-being. In 2011, Britain and France failed to deliver any of these goals in Libya because they lacked the forces, resources, campaign design and strategic weight and patience to deliver any of them. These basic realities should have been apparent to the Cameron and Sarkozy governments, both of which were engaged in Afghanistan.

A safe and secure environment is defined as the "ability of the people to conduct their daily lives without fear of systematic or large-scale violence." In 2024, much of Libya remains effectively lawless with public order virtually non-existent. The physical challenge alone is daunting. Even though 90 per cent of Libya's 6.5 million people live on the coastal strip, the country is roughly nine times the size of Scotland, covering 679,358 square miles or 1.7 million square kilometres of territory. The coast alone is 1,099 miles/1,800 kilometres long with land borders totalling

2,723 miles/5,000 kilometres. Furthermore, with a society built on tribes, clans and militias, establishing any legitimate state monopoly over the country was always going to be a tall order for two European former colonial powers with limited resources and limited backing from allies and partners. It has proven nigh on impossible and has led to Libya becoming one of the main conduits for the flow of irregular migrants from Africa to Europe.

Rule of law is defined as the "ability of the people to have equal access to just laws and a trusted system of justice that holds all persons accountable, protects their human rights and ensures their safety and security." A just legal framework for the whole of Libya is in 2024 yet to be established. Islamist groups insist on a strict interpretation of Sharia law, a position that led Berber representatives to walk out of several meetings held to discuss transitional arrangements. Public order, another key facet of rule of law, remains fragmented and uncertain. The equitable rule of national law is not yet established in Libya's two major population centres, Tripoli and Benghazi, and there is little chance Tripoli can expand its writ across the country. Furthermore, accountability under the law, access to justice, and a culture of lawfulness will likely require an entirely new system for the administration of justice.

Stable governance is defined as the "ability of the people to share access or compete for power through non-violent political processes and to enjoy the collective benefits and services of the state." Libya is still in post-conflict transition, which requires the steady and sustained reduction of conflict across security, economic, and political spheres. Tripoli is also only taking the most tentative steps towards representative government, with a hybrid political structure emerging of secular, tribal, and Islamist elements all vying for supreme state authority. How this equilibrium is institutionalised with the checks and balances in place to ensure that no single group dominates is perhaps Libya's most pressing challenge.

A sustainable economy is defined as the "ability of the people to pursue opportunities for livelihoods within a system of economic governance bound by law." According to the UN Development Programme Human Development Index, Libya ranked fifty-third out of 169 states prior to the civil war and enjoyed a relatively educated population with enough of a middle class to, in principle at least, provide an entrepreneurial impetus for the economy. One of the first-order requirements was to re-establish macroeconomic stability in areas such as consumer price inflation, gross domestic product growth, measured unemployment and employment, fluctuations in government finances, and currency. However, Tripoli lacks any real instruments over its conflict-torn economy and is thus incapable of establishing the functioning structures critical for effective economic governance. London and Paris, engaged as they were in Afghanistan, did little to assist.

Social well-being is defined as the "ability of the people to be free from want of basic needs and to coexist peacefully in communities with opportunities for advancement." Libya's greatest assets (apart from its people) are its high-grade hydrocarbon and gas reserves. These could in time fund the resources for meeting the basic needs of the people, but only in time. Tripoli has moved to establish new contracts with potential partners, and the post-conflict vacuum has enabled China, Russia, and Turkey to interfere in Libya's internal affairs.

Syria

Britain's Syria strategy had a beginning, a muddle, but no end. In Libya, Britain and France took military action when more prudence was required; in Syria Britain baulked at military action when the need was clearer and more legitimate due to the use of chemical weapons by the Assad regime. Consequently, Syria became a geopolitical vacuum which Putin's Russia moved to fill, emphasising Britain and the West's incompetence after repeated violations by Assad of President Obama's invisible "red lines," and the August 2013 UK parliamentary vote to prevent the Cameron government taking military action with the US and France. By 2013, burned by its experience in Afghanistan, Iraq and Libya, rather than learn the lessons from previous campaigns and adjust statecraft accordingly London simply lost its nerve. Even though Syria continues to threaten not just itself but the stability of the Levant and much of the Middle East, Britain and the West have shown themselves powerless.

In 2013, British Prime Minister David Cameron produced a thirty-six-page dossier (ghosts of Tony Blair and Iraq?) to convince parliament that the RAF could attack Islamic State beyond the non-existent Iraq-Syria border. In fact, all the subsequent debate achieved was to underscore just how fragile Britain and its foreign and security policy had become. Cameron's dossier revealed the lack of anything one might consider a comprehensive Western political or military, offensive or defensive strategy that could remotely deal with such a major threat.

The same failings that afflicted Britain's Libya policy were also apparent in Syria, most notably a lack of joined-up strategic thinking and a tendency to focus on inputs rather than outcomes. David Richards attempted to persuade both Cameron and Obama to create a credible Arab anti-Assad force when the Syrian dictator was at his weakest. The plan envisaged the raising, equipping, training and expert deployment under American and British command of an Arab Syrian force of around 100,000 in strength. When the idea reached Washington Richards was told that Obama's view was that it was more than the political market would bear in terms of US commitment and coin.

Syria has been riven by ethnic tensions ever since the minority Alawite community seized power in Damascus through President

Assad's father in 1966. Syria is 90% Arab, with some 2 million Kurds plus other smaller groups making up the balance of a population of 22 million, which grew some eightfold between 1950 and 2024, a demographic surge reflected across much of the Middle East. Ethnic tensions reinforce the sectarian divisions which helped Islamic State grow rapidly. 74% of Syrians are Sunni Muslims, 13% Shia, with the rest comprised of small Christian, Druze, and other communities. Until the war, the Ba'athist constitution protected minorities, and until those minorities again feel secure peace is unlikely to be re-secured.

The Syrian Civil War was and is a threat to the many precarious states in the region. In the wake of the Arab Spring (perhaps the less catchy but more accurate title would be the emaciation of the Arab state) these precarious states, such as Egypt, Iraq, Jordan, and Lebanon, all became more unstable precisely because all had to contend with a potent mix of contending ethnicities, sectarianism, economic decline, and enduring political tensions between states, rulers, and peoples. Through the spread of both al-Qaeda and Islamic State affiliates, such precariousness is extending to sub-Saharan Africa and the Horn of Africa. There is also the very real chance that nuclear-tipped Israel could be dragged in, especially if Jordan's collapse were to threaten some form of new Intifada that further exacerbates tensions between Israelis and Palestinians and sees Iranian-backed Hezbollah and Hamas join forces. Post 7/10 anything is possible.

The Syrian Civil War has also had profound implications for Europe through the mass irregular migration of millions of Syrians, Iraqis, Iranians, and others. The 13 November 2015 terrorist attack on Paris was an extension of the sectarian divisions that have riven Syria and much of the Middle East. Radicalisation of members of the many North African, Sub-Saharan African, and South Asian diasporas that now live in Europe means that a very small minority of people now pose a very real threat to Europeans of all ethnicities and beliefs.

The Syrian Civil War is also a potential flashpoint for growing geopolitical tensions. Despite being embroiled in Ukraine Moscow still has a free hand to strike anti-Assad forces whenever it chooses under the guise of its own war on "terror." Russia is also enabling Iran to extend its influence at the expense of the West in return for Iranian drones being supplied to Russia. This influence is all too apparent in Iran's support for Houthi forces in Yemen and the attacks on shipping in the Bab-al-Mandab. Turkey is also a regional and inter-regional player and has moved closer to Russia following the West's weakness in facing down Moscow. This is in part because Russia was resolute and determined and gained much kudos and influence with autocratic rulers of all persuasions in the Middle East. Perversely, when it came to confronting Islamic State Russia cooperated with the West because such cooperation was clearly in the Russian interest. Unlike Britain, Russia does not confuse values with interests. Consequently, Russia's stock in the region is still much higher than the West should find comfortable and indeed wishes to acknowledge.

What all the above reveals is again the essential tension between the ends, ways, and means of British strategy and a tendency to grandstand on the basis that "something must be done," whilst having no clear understanding of the British interest and lacking the influence and the means to shape anything but the most marginal of outcomes. This is dangerous, and Britain could play a much more pragmatic role in the Middle East than it does. The Syrian Civil War remains the epicentre of conflicts across the Middle East that are now breaking out of one region and beginning to destabilise others. The destruction of Gaza is but the latest example.

Whilst the time for a pan-Arab force is over, a comprehensive strategy is still needed that works at the ethnic, sectarian, regional, inter-regional and geopolitical levels. Britain could help lead the crafting of such a strategy, built on the principle of Western states working in close partnership with states in the 7/10 and Israel's assault on Gaza means there can be no peace in the Levant for the foreseeable future. Gaza is the latest example of conflict creep. But the so-called Abraham Accords agreed at the White House in 2020 could in time form the basis for a Western-backed anti-Iranian strategy that would move more Gulf States to recognise Israel (Bahrain and the United Arab Emirates already do), and if properly managed, thus help move the Palestinians far closer to a peace that would see Hamas and Hezbollah excluded. Given the importance of the region to European security (look at a map), and the number of refugees who have entered Europe because of the conflict creep, Britain must use its good offices to help forge American and European strategic and political leadership. After all, Americans and Europeans have every right to involve themselves given the threats to their critical interests, but at present there is only vaccilation. The West also needs to remind Moscow where real power lies following Moscow's successful humiliation of the West and its leaders, who were reduced to impotent political hand-wringing over Syria. Inaction has consequences. Not only did President Putin seize the agenda, his success clearly emboldened him to seize Crimea and Eastern Ukraine.

Afghanistan

What happened in Afghanistan was a monumental grand strategic failure for the US, Britain, and their Western allies. The US was humiliated, as was Britain, whilst other Europeans were revealed as security lightweights way out of their depth in a country in which they really did not want to be. As such, the campaign in Afghanistan is perhaps the greatest example of the unintended consequences of strategic virtue signalling and the confusion of values with interests. As the only non-American Commander of the post-2006 Afghanistan-wide International

Security Assistance Force (ISAF), and subsequently as the military head of Britain's armed forces, David Richards saw all these political failures close-up, as the last dying embers of Blair's liberal humanitarian interventionism were extinguished and a new age of realpolitik heralded in. There were simply too many actors with too many aims and too many rules of engagement. Worse, as the US became less emotional as 9/11 receded from immediate memory it became ever more unsure and unclear what it was trying to achieve in Afghanistan.

It was not the first time British forces had been in Afghanistan, but for most Britons, including its leaders, it was the most unexpected. The failure in Afghanistan emerged from the smouldering remains of the World Trade Center in New York, the broken Pentagon, and the fragments of Flight 93 in that epoch-making sunny early autumn morning in September 2001. Two months later British forces began arriving in Afghanistan to take on a mission for which they were ill-prepared and of which Prime Minister Blair and his team seemingly had little idea of the challenges.

It was not quite the First Afghan War redux, but the eventual 2021 victory of the Islamic Emirate of Afghanistan was not far short of it, and it might have helped if Blair and his team had paid more attention to British history. On 1 January 1842, following an "agreement" between Major-General William Elphinstone, commander of British forces in Afghanistan, and Pashtun warriors, the British began what they thought would be safe passage from Kabul to British India for 4,500 military personnel and 12,000 mainly Afghan and Indian camp followers. As soon as the retreat began the civilians and their military escort came under attack from Ghilji warriors. On the second day the Royal Afghan Army's 6th Regiment, in which the British had invested much effort, deserted. Ultimately, only one British officer and seven Indian soldiers survived the ensuing massacre.

Between 2001 and 2021 some 2,500 US military personnel were killed in Afghanistan, along with over 1,200 other Coalition personnel, of which 456 were British. More than 66,000 Afghan soldiers and police were also killed and many more no doubt since the Taliban retook power.[3] At least 47,000 Afghan civilians were also killed with almost 400,000 Afghans displaced since May 2021 alone, whilst over 50,000 Taliban fighters died in the conflict.[4] The cost of training and equipping the collapsed Afghan National Army is estimated at some £64bn whilst President Biden says the US spent over $1.5 trillion on the Afghan campaign.

During the worst times of the Troubles in Northern Ireland there were around ten British soldiers for every 1,000 citizens. In Afghanistan, there were around 0.8 troops per 1,000 citizens, which puts into perspective the British decision to move out of Kabul and the protection of the seat of government into Helmand Province, especially given that in Bosnia there were five troops per 1,000 citizens. Indeed, if the contemporary Bosnian model of peacekeeping had been adopted—peacekeeping in what was a relatively benign environment following the Dayton Accords—coalition forces would still have needed a force of at least 240,000.

It was also a catastrophic failure of high political leadership, and not just in Washington. All the tests of grand strategy were ignored with tragic consequences for the Afghan people that will last for years, not least the significant numbers of Afghans making their tortured way to British shores. After the suffering of the Afghan people because of the West's broken promises, the greatest loss has been to the reputation of American leadership and British competence. It is not at all clear that either Britain or America has the political capacity or strategic patience to stay the course of any prolonged campaign. The arbitrary decision the Trump administration made to leave Afghanistan, confirmed by President Biden, simply gifted the Taliban a victory that had far more to do with America's toxic domestic politics than considered US or Western security strategy, as evidenced by the unedifying blame game between presidents Trump and Biden. Britain? Is it really a good idea to tie so much of Britain's future strategic assumptions to those of the United States?

Ideally, Britain should also never again be so reliant on the US for so many of the critical decisions upon which it, and its military personnel, depend that options are effectively denied London. The failure in Afghanistan will affect Europe and Britain far more than America but the withdrawal was imposed on the allies by a decision taken in Washington with little heed for America's partners, and little care. Worse, in their haste to get out the Americans did a deal with the Taliban that has very significant security implications for Britain. Under the February 2020 Doha deal between the Trump administration and the Taliban, US forces were to have been withdrawn by 1 May 2021. In return, the Taliban would break links with al-Qaeda and enter peace negotiations with the Ghani government. They did neither. Worse, as the date for completion of the withdrawal slipped back to 31 August, it established a clear link in the minds of the Taliban leaders in Peshawar and Quetta between the withdrawal and the twentieth anniversary of 9/11. Given that information warfare and propaganda is a large part of what the Taliban and other Salafist Jihadists use to subvert democracies, the anniversary clearly spurred them on to retake Kabul by 11 September. This helped accelerate the collapse of demoralised and scared Afghan National Security Forces (ANSF), stripped of their foreign advisors and key air support in the face of an implacable enemy. Washington not only handed the Taliban an unconditional victory but afforded violent Salafist jihadis the world over an immense propaganda coup and a powerful recruiting tool. Jihadist groups across the world even interpreted the defeat of the West as the fulfilment of a prophesy that a Muslim army would defeat infidels in the so-called Khorasan, which includes parts of Afghanistan.

Did the American withdrawal make good grand strategic sense and did Britain agree? Justifying his decision, President Biden said,

> Today, the terrorist threat has metastasized well beyond Afghanistan: al Shabaab in Somalia, al Qaeda in the Arabian Peninsula, al-Nusra in Syria, ISIS attempting to create a caliphate in Syria and Iraq and establishing affiliates in multiple countries in Africa and Asia. These threats warrant our attention and our resources.

He also said, "There was only the cold reality of withdrawing our forces or escalating the conflict and sending thousands more American troops back."[5] Twenty years on from the December 2001 invasion Western powers were certainly right to be looking to Afghans to decide their own future—and in a terrifying way they have, or at least have had it decided for them. Some argued that the rise of China and the return of Great Power competition means it is simply no longer possible for the US and its allies and partners to commit such large parts of their respective forces and resources to one central Asian country. It is also true that the Taliban reduced its attacks on the Coalition in the wake of the February 2020 deal with Trump and that had the US reversed the decision to withdraw such attacks would likely have resumed.

The precise problem was neither military capability nor capacity, but a lack of grand strategic clarity allied to an absence of political will. By 2021, the US and its allies had a far better understanding of Afghanistan than twenty years prior, and this enabled them to settle on a strategy of support for a Western-backed ANSF. That has now been lost, not least because such support was removed with catastrophic consequences. In fact, the US and its allies had succeeded in striking a balance in which the number of Allied and Partner forces involved were relatively limited, and could have been sustained at relatively little cost for many years, buying time for a generational change in the population's outlook. With political will behind them they should have been able to both prepare for the challenges posed by the likes of China and Russia whilst maintaining the mission in Afghanistan.

A lack of strategic clarity was evident for much of the campaign, something which President Biden let slip when he said, "Our mission in Afghanistan was never supposed to be creating a unified, centralised democracy."[6] In fact, that was precisely the mission the US signed up to in December 2001 as part of the Bonn Agreement, as it was the price many Europeans demanded for committing to a long-term campaign in Afghanistan.[7] Back in December 2001, when Coalition forces entered Afghanistan, the West could have mounted a "simple" search and destroy counter-terrorism mission. However, the West (America and Britain included) chose not to and instead stayed and attempted to build a functioning Afghanistan that would no longer be a threat to itself or others, a firm promise that was also made to the Afghan people. Sadly, it is that promise that was betrayed, a betrayal witnessed the world over with the most profound strategic consequences.

The hard truth is that the West failed in Afghanistan because the Americans were never sure how to conduct such a campaign, and the British and others were never willing to commit enough force and resource to make it possible to meet the challenge from the outset. Presidents Obama, Trump, and Biden all signalled that they were far more concerned about getting out of Afghanistan than turning it into a functioning state. The real driving force was the impact on their electoral chances. Europeans? Most of them tried to limit the exposure of their young men and women from the moment they arrived. Only the British, and to some degree the Canadians, over-reached: Britain was stuck between wanting to give the impression that it could do far more than the other European allies and being dependent on the Americans in order to make this appearance possible. It was strategic and political folly for which 456 British and 158 Canadian servicemen and women paid with their lives.

The Afghanistan debacle is thus a tragic example of what happens when grand strategic ends, ways and means get out of sync. In 2009, David Richards, then-head of the British Army and fresh from his tour as Commander, International Security Assistance Force (COMISAF), was interviewed by the BBC. He suggested that it could take another thirty years to establish the conditions in Afghanistan sought by the international community. At the time, Richards' assessment was widely ridiculed. He was correct. Such stability rested on efforts by the Coalition to build governing institutions sufficiently respected and capable of governing the country for the good of all Afghans. On a visit to Kandahar and southern Afghanistan in 2008 it was obvious to Julian Lindley-French that such an aim was a herculean task beyond the means afforded those in command, not least because their efforts were frustrated by corruption at the very highest levels of government. President Biden said that Americans were tired of "forever wars" and that Afghanistan is not the threat it was (at least for the moment). He also said that forging competent governance in Afghanistan is not possible, which if correct was an admission that the entire campaign design and strategy had been deeply flawed from the outset.

Throughout the campaign commanders grappled with a deadly paradox: many leaders in Washington did not believe in state-building, whilst many leaders in Europe who insisted on state-building signally failed to invest in it. Britain found itself trapped between those two positions with the Blair, Brown, and Cameron governments routinely uttering platitudes which they did not believe and which could not be achieved given the intelligence and lack of resources. Consequently, the counter-terrorism campaign against al-Qaeda and the stabilisation campaigns too often ran in parallel and even came into conflict despite efforts by commanders in the field.

The crucial failure was the extension across the country of the writ of the Kabul government, which many Hazara, Pashtun, Tadjik and Uzbek Afghanis alike saw as corrupt and incompetent. Worse, the Allied campaign was poorly designed, fractured, and unevenly spread across a host of so-called "provincial reconstruction teams," all of which were different and more or less competent depending on which country was responsible. David Richards attempted to bring cohesion and unity of purpose to the overall campaign through the creation in 2006 of Afghan Development Zones (ADZs) overseen by a new campaign-directing council called the Policy Action Group, chaired personally by President Karzai, who strongly supported the ADZ concept. Sadly, it did not long survive Richards' departure from Kabul in 2007.

The Collapse

The final collapse of the Afghan National Army was a lesson in the price of grand strategic hubris. On 8 July 2021 President Biden said,

> [T]he Afghan troops have 300,000 well-equipped [personnel]—as well equipped as any army in the world—and an air force against something like 75,000 Taliban ... The Taliban is not the North Vietnamese Army. They're not remotely comparable in terms of capability. There's going to be no circumstance where you see people being lifted off the roof of the embassy of the United States from Afghanistan.[8]

In fact, without the core pillar provided by the US and its allies the Afghan Army was always a house of cards. From the outset the campaign had been predicated on the belief that over time credible Afghan National Security Forces could be fashioned, and a shared Afghan identity emerge sufficiently robust to provide effective pan-Afghanistan security. It was the plan of the late Donald Rumsfeld, and it was that plan which failed so catastrophically. NATO Secretary-General Jens Stoltenberg's claim that the collapse of the Afghan government was due to the failure of the Afghan National Army (ANA) is correct in a very narrow sense, but the real reasons the ANA collapsed are far more complex.

Whilst the ANA had some 30,000 excellent Special Forces much of the rest of the ANA was an immature force, with its *kandaks* (battalions) organised around and dependent upon US command and strategic enablers, most notably air power, technology, planning, and logistics. On 4 July 2021, much of that US core was withdrawn suddenly and precipitately over a twenty-four-hour period. Unsurprisingly, there was a catastrophic collapse in morale, without which even the best armies will not operate effectively. Contrary to what President Biden said, the real strength of the Afghan National Army was also nothing like 300,000, with many so-called "ghost soldiers" who simply did not exist, although their pay was claimed by their commanders. Much of the force that did exist was poorly fed and even more poorly led, whilst their fuel and ammunition were sold off on the black market before reaching them. Therefore, it is hardly surprising that much of the ANA melted away as the Taliban marched across Afghanistan. Only the 201st Corps and the 111th Capital Division stood their ground, and tragically they were destroyed. In other words, the "ANA" was only ever going to be credible as a force if the US and its allies remained in country to support them. Today, many of the weapons systems supplied by the US and Britain to the ANA are now on the global black market.

In the end Afghanistan, Iraq, Libya, and Syria found out Western strategy. The geopolitical and even societal implications for Britain will be profound. The Americans will refocus their attention on warfighting and "pivot" (to use that ghastly phrase) towards China and preparations for some hi-tech, high-end robotic future war. The British will go with the Americans as far as they can, although the British will also claim they have gone far further than they have, just as the British always do.

The future? The autocracies and democracies are in a strategic competition to woo the Global South. As Kabul was falling another event was taking place that was rich in grand strategic resonance. Seventy-five years to the day after Indian independence, India's INS *Tabar* ("Battleaxe") sailed into Portsmouth, the fleet headquarters of the Royal Navy. The grand strategic implications were clear. Democracies the world over face a growing range of threats and to maintain the peace nothing short of a new concept of grand strategic multilateralism must now be forged and Britain must help to forge it. However, any such grand strategy will only work if the Global South believes the West has learnt its lessons about the need for greater strategic modesty and the need to listen to the concerns of the Global South.

The West will continue to have more watches than time (to paraphrase that well-worn Afghan aphorism that also turned out to be a truism) but unless it learns again to have strategic patience and match ends and ways with means then there will be more Afghanistans, more Iraqs, Libyas, and Syrias, and far more China and Russia. There were certainly failures and mistakes made by commanders in the field. Mistakes that were inevitable given the complex nature of the place and the mission. However, ultimate responsibility for these failures must rest with those political leaders who willed the ends without clear ways and with critically inadequate means. The men and women charged with the conduct of these missions by their political masters did their utmost to make flawed strategy and policy work, but they were let down by their respective capitals trying to close a political gap which was not of the soldier's making and for which many paid with their lives. Ultimately, the disaster in Afghanistan is due to a catastrophic failure of political leadership.

In August 1842, British Indian forces under General Pollock returned to Afghanistan and inflicted a massive defeat on the Ghilji as revenge for the destruction of the British column a year earlier. In September 1842, the British re-entered Kabul and captured Dost Mohammed Khan, one of Afghanistan's most powerful tribal leaders and the first commander of what might be called the Afghan Army. He asked his British captors a question. "I have been struck by the magnitude of your resources, your ships, your arsenals, but what I cannot understand is why the rulers of so vast and flourishing an empire should have gone across the Indus to deprive me of my poor and barren country."[9] Good question.

Gaza

Effective grand strategy must be driven by contemporary national interests, not some vague sense of historic guilt. Such strategy also demands that at times London must be tough on friends who damage their own interests and Britain's. UNSCR 2728 of 10 June 2024 may have reduced the slaughter but there will be no peace in the Levant anytime soon. The war in Gaza is thus not just a test for the Israelis, but also for Britain.

Britain has been directly involved in the Levant since World War One. On 2 November 1917, the Balfour Declaration committed the British government in principle to support the setting up of a Jewish state in what was then Palestine, part of the failing Ottoman Empire. The creation of the State of Israel in 1948 was always going to create tensions across the Middle East, much like the unification of Germany at the heart of

Europe in February 1871. Few predicted the enduring and tragic nature of the struggle that has ensued and ever since British policy has been trapped between its strategic interests in the region and its values. That enduring tension is apparent in London's uncertain response to Israel's heavy-handed use of firepower in the wake of the 7 October 2023 massacre by Hamas of over 1,200 Israelis, some in the most barbaric way imaginable. The consequence of almost unquestioning support for Israeli actions has ceded the strategic narrative to China and Russia across large parts of the Middle East and the Global South. This is driven not just by Britain's fawnlike relationship with the US, but also by how difficult London finds it to distinguish between support for the Jewish people and support for Zionism. Despite the Israeli Defence Force's (IDF) clear breaches of international humanitarian law London has clung doggedly to supporting the Israeli position and thus offered little benefit to either. Paradoxically, the tragedy of 7/10 and Gaza is one of those occasions when interests and values should align. Britain should stand up and remind all parties to the conflict that playing into the hands of the Iranians makes little or no strategic sense.

Do what your enemy least wants is a dictum of war that has endured from Lao Tzu and Sun Tzu to Machiavelli and Clausewitz. The 10/7 Hamas terrorist atrocity carried out by the Iranian-trained Nukhba force was an outrage. Hamas has proved again they are as much a curse for the Palestinian people as they are for the Israelis. Israel, Hamas, and Iran are engaged in a war of existence, even if it is often a proxy war of existence.

Today, Israel finds itself in a similar place to where the Americans were in the immediate aftermath of 9/11—trapped between anger and strategy. Back in 2014, Israel launched Operation Protective Edge and, just as it did again in November 2023, invaded Gaza, one of the most densely packed urban environments on Earth. With its range of armoured vehicles including Merkava main battle tanks, armoured personnel carriers and D9 bulldozers, the idea in 2023–4 was to minimise casualties amongst the IDF—but not only were over 500 Israelis killed (latest UNOCHA figures) civilian casualties have been horrendous. In February the United Nations Office for the Coordination of Humanitarian Affairs (UNOCHA) estimated over 28,000 Palestinians had been killed, with almost 68,000 injured.[10]

Israel is playing into Iran's hands and in so doing Tel Aviv risks undermining the chances of Saudi Arabia and the Gulf states recognising Israel, beyond Bahrain and the UAE. As a friend of both the Gulf states and Israel, London should be reminding Tel Aviv of that. Much of the world is also turning against Israel. Naturally, like the Americans post-9/11, the Israeli need for vengeance for 10/7 is overwhelming. When British cities were bombed by the Luftwaffe in 1940 and 1941 few in Britain questioned the RAF striking back at German cities and civilians, but neither response made for great warfighting strategy. Specifically, London needs to remind the Israelis of the bigger strategic issue at stake. The reason for the 10/7 attacks on Israel is different to previous attacks by both Hamas and Hezbollah. Israel already has an effective working relationship with Egypt and is also close to a regional-strategic rapprochement with the Saudis, with profound implications for Israelis and the wider Middle East. If such a relationship can be secured it will further isolate Iran and by extension Hamas and Hezbollah.

Despite Israel's understandable anger, the political damage done to Prime Minister Netanyahu, and the Israeli tradition of an "iron fist" response to all and any such attacks, it is not in Israel's interest—beyond the profound legal and ethical considerations—to again kill large numbers of Palestinians or to deny them food, water, and medical treatment. That is precisely what Iran and Hamas want them to do. Rather, Israel should rebuild the defences on its southern border with Gaza and go after the Hamas leadership over time and space in that time-honoured and highly effective Israeli way. The best way to defy Iran and Hamas is to build on Tel Aviv's relationships with Riyadh and Cairo (and listen to Egyptian intelligence), and by so doing preserve the sympathy of those who support Israel's right to exist and defend itself proportionately. In other words, Israel will only succeed in this struggle if it does what Hamas and Iran do not believe Tel Aviv will or can attempt: revert to a proportionate and merciful response that respects the constraints of international humanitarian law.

If Britain does not want to be dragged repeatedly into conflicts in the region the only longer-term strategy is some form of two-state solution, made much harder by illegal Israeli settlements in the West Bank. The creation of a Palestinian state would crucially also take the ideological legs from under Iran, Hamas, and Hezbollah. Both Britain and the US need to be very robust with Israel over this issue, placing British, and frankly Israeli, interests above any ill-thought-through endorsement of extreme and illegal action by Tel Aviv.

The Tests of British Action

There is one compelling lesson of the recent "something must be done..." era of British power. Starting in the wake of the Cold War and the wars of the Yugoslav succession which began in December 1991, unless a vital British interest is at stake London's political will and materiel support very quickly erodes. The greater the ends that are apparently aspired to "over there," the weaker London's political will and British means "over here."

Ever since the Treaty of Westphalia in 1648 created the nation-state and began to move power away from capricious aristocracies a link was established between interests and proportionality of action. A vital or critical interest leaves a state with no choice other than fight or submit, a choice Ukraine now faces. Essential interests, whilst important to a state, leave room for chosen emphasis across the diplomatic, informational, military, and economic instruments of power. General interests are more values-led and tend to be the stuff of hegemonic power in environments where no other state or group of states can realistically counter the export of values through power.

What about Britain's recent campaigns in Afghanistan, Iraq, Libya, Syria, and Ukraine? Sound strategy not only fosters clarity of action, but it also helps establish criteria for action. This is what Tony Blair tried to do in his Chicago speech as he attempted to strike a new balance between vital, essential, and general interests in what in 1999 was by historical standards a relatively permissive geopolitical environment. Today is different. In that light, was the case for using British military power in the recent campaigns clear and had all diplomatic channels and options been exhausted? Were military operations undertaken prudently, with a reasonable chance of securing their political and humanitarian goals? Was Britain prepared for the longer-term consequences of such action? Were there clear national interests at stake?

The 2001 invasion of Afghanistan was probably inevitable in the wake of the massive loss of life, including British life, on 9/11. The Americans were in no mood for diplomatic options and had Britain said no to Washington's desire for revenge it could well have been the end of NATO. This is why for the first time in its history the Alliance invoked Article 5 of the Washington Treaty on 12 September 2001. Thereafter, there seems little relationship between the military and other means Britain deployed to Afghanistan and the political and humanitarian aims for which such forces and resources were committed. The longer-term consequences included the destabilisation of British cities and towns where growing Muslim communities existed. There were certainly no national existential interests at stake.

The 2003 invasion of Iraq was for Britain partly a consequence of 9/11 and partly again the need to follow the Americans. In 1998, the Good Friday Agreement ended thirty years of sectarian violence in Northern Ireland. When President George W. Bush invoked the Global War on Terror in the wake of 9/11 London seized the opportunity to end American support for the Provisional IRA by linking them with al-Qaeda. At a stroke, the IRA's armed struggle was ended. However, when Bush linked Saddam Hussein to the "Global War on Terrorism," Prime Minister Blair had no option other than to support Washington even though British troops were already engaged in Afghanistan. London had become hostage to the fortunes of Washington neo-conservatives who really saw the unseating of Saddam as part of a new Crusade for freedom. At no point during the post-2003 campaign in Iraq did Britain have sufficient influence, capabilities, or capacities to realise any of the political and humanitarian goals it claimed to have. Rather, Britain was an American auxiliary with strategy made elsewhere.

The one occasion London sought to act autonomously from the Americans was the joint attack on Libya with the French in 2011. The stated aim was to prevent a genocide by the Gaddafi regime in Benghazi, but that fear was based on faulty intelligence. What happened thereafter had echoes of the 1956 Suez fiasco. It soon became clear that without the support of American "enablers" and given the ongoing commitment in Afghanistan and the attrition from which British forces suffered after six years of hard engagement in Iraq between 2003 and 2009, the most London and Paris could do was to assist the Libyan opposition to topple Gaddafi. Or, in other words, foster chaos. Political and humanitarian goals?

Syria was sidestepped by undertaking only punitive strikes against Islamic State, something which the Russians exploited in September 2015 by using the pretence of the war on terror to effectively seize control of Damascus. The consequences? David Richards is unequivocal: Syria typified Britain's confusion of values with interest. London did far more than simply act against the Salafist Jihadis of Islamic State, the British also gave a lot of indirect military and moral support to anti-Assad forces, encouraging them with all sorts of promises, as London had other groups in Afghanistan, Iraq, and Libya. Much like Ukraine, London gave the Syrian opposition just enough to keep fighting but never enough to give them a chance of toppling Assad. The result was (and is) a very long, drawn-out conflict in which over 800,000 people have lost their lives with millions more displaced, of whom a significant percentage have made their way to Western Europe and Britain. Richards spoke to anti-Assad commanders who say openly that if they had known in 2011 how inadequate in practice Western support for their cause was going to be, they would not have revolted against Assad. London's self-deceiving excuse is to quietly blame the Americans, where much of the blame must indeed lie, but Britain too must take responsibility for its recent campaigns being a shameful strategic and moral failure.

Well-meaning and gung-ho at the outset when the "something must be done" lobby is in the ascendant, politicians give military commanders rules of engagement, forces, and resources to just about keep a campaign going, but never enough to prevail. The same pattern of pretence is evident in Ukraine where all Britain (and its partners) are doing is to keep the Ukrainians in the fight, but with little or no chance of expelling the Russians from their land, or even returning Ukraine back to its 2014 borders. Richards is clear: President Zelensky's instinct was to negotiate with Putin in the month following the February 2022 invasion, but Boris Johnson and others persuaded him to fight on with a host of promises of Western support that only partially materialised. Once again, London demonstrated not only a failure to understand Russia, but also ignorance about the physics of war.

The only campaign which did meet Tony Blair's criteria for action was Sierra Leone in 2000. The case for British military power was clear because of the impending threat to British and other citizens in the capital, Freetown. There were no diplomatic channels to the RUF and its leader Foday Sankoh. The military operations undertaken were both prudent and had a reasonable chance of securing their political and humanitarian goals. There were few if any negative longer-term consequences of action, other than perhaps hubris. Whilst there may have been no existential national interests at stake, Sierra Leone was doable at a time when the West was still sufficiently dominant to "do." That is no longer the case. The lessons? Speak softly and carry a biggish stick. Do not speak loudly and often if one only carries a small stick. In other words, if Britain is going to go to war, in whatever form it takes, do not do it with a peacetime mindset. War is war! If you choose to fight it, you must choose to win it!

3

SMOKE AND ERRORS

Nobody could say that from any one moment war was an impossibility for the next ten years ... we could not rest in a state of unpreparedness on such an assumption by anybody. To suggest that we could be nine and a half years away from preparedness would be a most dangerous suggestion.

Arthur Balfour, British Prime Minister, 1932

The Virtual Ten-Year Rule

What are the consequences of national strategic pretence? The terrible titans of the post-COVID world are geopolitics and geo-economics, neither of which are trending in Britain's favour. The world is witnessing a profound shift in the balance of coercive power away from the democracies towards China, and by extension its piggyback partner, Russia. This shift in the global balance of coercive power has the most profound implications for Britain, but that is not the impression given by London.

Rather, British security and defence strategy is built on a series of assumptions about the scale and nature of threats,

Britain's ability to do something about them, and the role of allies and partners. The core assumption behind British strategy is that America will always be there, be it as a shoulder for Britain to stand on, or a saviour upon which Britain can rely. This is at best optimistic. And the price of British pusillanimity? The abandonment of substantive and sound British strategic autonomy in favour of defence pretence. To understand that contemporary truism of British strategy one also must understand why Britain has retreated so quickly from grand strategy and the consequences for British defence policy. It is a retreat that began in 1967 with the withdrawal of British forces from east of Suez and the establishment of a culture of decline-management that continues to this day.

Strategic Defence Review 1998 crossed a Rubicon by breaking the vital link between policy and planning. At a stroke, it went from being a classified driver of policy to an exercise in public relations and strategic communications that saw spin begin to replace substance at the heart of British national and defence strategy. Effective strategic planning requires facilitators trained not only in the art of strategic assessment, but also the science of planning, coordination, and implementation, reinforced by enough expertise within government to assist in strategic processes, and supported by a bureaucracy self-confident enough to welcome the testing of its assumptions. Such "red teaming" is vital across the conflict spectrum and for the reasoned generation and application of national diplomatic, informational, military, and economic power. Such expertise may exist within ministries, but if not, should be readily available. The critical issue is whether it is listened to. The late Professor Sir Michael Howard, one-time Oxford tutor of Julian Lindley-French, believed historians made the best strategic planning experts. This was because of his belief in the need to gain both wide and deep expertise across policy to better understand the interaction of politics, socio-economics, culture, tactics, and strategy.

In August 1919, at the behest of the then-Secretary of State for War and Air Winston Churchill, Britain adopted the so-called Ten-Year Rule. Under the rule, London assumed that it would not be engaged in a major war for at least a decade and could thus cut defence spending accordingly. In March 1932, shortly before the rise of Hitler in Germany, Britain scrapped the Ten-Year Rule. In 1934, following the collapse of the Geneva-based World Disarmament Conference, and the defenestration of the League of Nations, Britain embarked on a massive military rearmament programme which helped it narrowly avert defeat in 1940. Today, Britain is trapped in a kind of virtual Ten-Year Rule that affords London the comforting blanket of false security even though all the evidence is that another major war might be far more imminent than Britain's leaders would like to think. The British cannot be safe, nor will they make their own region or the wider world safer, if they continue to hide in the virtual world to justify London's decision to accept greater risk by being intentionally too weak for a country of Britain's wealth and power. This issue goes to the heart of Britain's problem with strategy: what power and kind of power would Britain need to generate if London is to keep Britain secure and prevent war? What should be the aims of a contemporary British national security strategy given Britain's critical interests and its place in the world?

Former British Prime Minister Tony Blair is worth quoting at some length on this question.

> Strategy would begin by an analysis of Britain's strengths. It has a strong alliance with the USA. It is part of the continent of Europe which contains the world's largest commercial market and political union. It has the global links of the Commonwealth, which, carefully nurtured, can be modernised, and exploited. It has the English language. A world-class capital city. Highly respected Armed Forces. Fantastic life sciences. World beating Universities. A highly

> developed technology sector, particularly now in AI. It has (or had) the world's premier financial centre. It also has renowned arts and culture. The right strategy would start with how each of these strengths is built into something globally meaningful. My government had three legs to the foreign policy stool. We would be the strongest ally of the USA, their first call across the Atlantic. We would be leading players in Europe, capable where necessary, of blocking the Franco-German motor and would be the Americans' chief friend in Europe. And we had in our aid and development commitment—symbolised by DFID [the Department for International Development]—the best soft power instrument in the world. And our Armed forces could compete with anyone not in size but in quality and fighting power.[1]

London cannot do all these things to effect: all the recent defence reviews have fudged them at best. The first photograph in the main body of Integrated Review Refresh 2023 shows two impoverished women in a developing country. Their well-being is of course important, but what the photograph reveals is how London sees the world. First, global development is more important than national defence. Second, the security of the British citizen is inextricably linked with the two women photographed and, consequently, "solving" global poverty is at least as important as defending British citizens. In other words, at the core of Britain's paramount security strategy document is a globalist paradigm that regards the nation-state with some distaste.

London would object to the above statement by pointing to the "temporary" cut in the development budget from 0.7% GDP to 0.5%, whilst also pointing out the aspiration or commitment to increase the defence budget to 2.5% GDP, albeit by 2030. However, whilst the relationship between the national interest, the defence budget and those who pay for it (the taxpayers) is clearly demonstrable, there is too often little relationship between the national interest and the development budget, which can appear little more than a post-imperial, virtue-signalling, globalist reflex. The strength of this post-imperial globalist belief system is such that London, at best, seems locked in a post-Cold War worldview that belongs to the 1990s. Time and threat have moved on. In other words, London suffers from a complex mix of hubris and guilt.

Leadership Versus Management

Mick Ryan states,

> Most political and bureaucratic entities in the West are now loath to take risks—the incentives are not there from the electorate to do so (or so they believe). Therefore, where strategy does exist, it is measured, bland and does not include anything that might negatively impact on a government—or the civil servants that advise it—down the track. This is not an issue of resources. It is an issue of leadership, strategic risk management and creativity.[2]

Such risk-aversion is reinforced by Whitehall group think and often toxic bureaucratic politics, partly caused by a culture that emphasises conformity rather than agility. For London "strategy" is also a battlefield in which the eternal struggle over money is fought between over-mighty ministries often with contending views and needs. That such a situation obtains at all reveals a lack of consistent strategic leadership from No. 10 Downing Street (and far too much from No. 11) as well as a weak, insubstantial, and inconsequential National Security Council that simply does not carry sufficient political or bureaucratic weight to drive synergies across government if ends, ways, and means are to be routinely balanced.

In the power pecking order of bureaucratic politics HM Treasury is the Alpha (the Prime Minister is, after all, also the First Lord of the Treasury), followed by the Home Office, Health,

Education and Justice. Whilst the Foreign, Commonwealth and Development Office is afforded significantly more importance than the Ministry of Defence, it is still well down the order of priority. One has only to see the place of defence in a Cabinet agenda to see how little importance it is afforded outside of crisis situations. This is also why London tends to exaggerate British soft power whilst avoiding hard power and any net assessment of Britain's critical interests that would drive defence priorities in a quantifiable and actionable manner. Contrast this with the Americans. The US Office of Net Assessment sits close to the Department of Defence and has always recruited regional and topical expertise to develop its understanding of adversaries and to develop counterstrategies, with much of the effort focused on non-military instruments of national power to counter major state threats. Such a long-term/permanent assessment capability could go a long way to helping Britain better conceive and implement security and defence strategy.

Because there is no grand strategic clarity at the top of government there is very little understanding of the utility of British power beyond muddling through. And because the focus tends to be value- rather than interest-driven there are also few mechanisms for driving integration at times of crisis. The COBRA (Cabinet Office Briefing Rooms) crisis management structure, such as it exists, emphasises collective rather than common action. Worse, during what is a systemic struggle there is nothing that conveys a sense of urgency for dealing with the threats identified. Consequently, policy responses are incremental and whilst they pretend to be threat-led they are in fact resource-led.

Nor is Downing Street the White House. A relatively weak Downing Street is "supported" by an often dysfunctional and weak Cabinet Office. Former Secretary of State for the Cabinet Office Michael Gove told the COVID Inquiry in November 2023 that (as the *Guardian* puts it), "the Cabinet Office, the coordinating department in government (of which he took charge in February 2020) became dysfunctional over time in a "piecemeal and cumulative way" by having more and more responsibilities added.[3] But it got little added authority. Like medieval barons under a weak king the COVID pandemic also revealed a lack of any robust unity of purpose and effort. Too often one arm of government seemed (or wanted) not to know what another arm was doing. These weaknesses are also exploited by the legal state as judges and other quasi autonomous agencies increasingly insert themselves into what was once the business of government by instrumentalising the Equality Acts, Human Rights Act, and the European Court of Human Rights (ECHR).

The very risk-averse nature of government is designed to avoid making tough choices and hard decisions. Risk-aversion is nowhere more apparent than over the torrid question of legal and illegal immigration. For many years, government has promised to markedly reduce immigration, whilst an aging population, a failing education system, universities, and the business demand for cheap labour in both health and social care have driven a marked increase in net immigration. Moreover, HM Treasury sees mass immigration as a means of boosting the nominal size of the economy, even as GDP per capita falls. Repeated governments have lacked the necessary political courage to deal with such a challenge, as in so many areas, with the resultant decline in social and political cohesion evident today. As Lionel Shriver wrote in the *Spectator*,

> according to newly released ONS data, net legal immigration is hitting record levels—revised upwards to 745,000 from 606,000 for 2022 and 672,000 for this year [2023] (not yet revised upwards). That "net" gambit, too, is an intentional obfuscation of Britain's fast-forward ethnic make-over. Of the roughly 500,000 emigrants who left Britain last year, only about 200,000 were neither British nor EU citizens. Of the 1.2 million immigrants who settled here last

> year, then, roughly a million were non-EU arrivals, who on average extract more than they contribute to the state.[4]

Equally, there are specific areas of national life where Shriver must be challenged and which advise against the blanket application of statistics on immigration. For example, 10% of all British Army personnel come from Commonwealth countries, and the National Health Service and care sector is hugely dependent on them.

The nature of the debate over immigration is often simplistic and too often reflects the increasingly toxic nature of political and public discourse. However, for a state of some 68 million people to absorb a gross (as opposed to net) inflow of over a million incomers every year without adequate investment in public services and housing is simply poor planning. Using mass immigration to doctor the size of Britain's GDP by driving ever more of the population into relative poverty though reduced GDP per capita is self-defeating. Why does it matter to security and defence? Sound defence can only work if it is part of an implicit contract with the people. Put simply, no state the size of Britain can cope with such inflows year on year and maintain the political and social cohesion vital to any foreign, security or defence policy. Shared identity matters.

London is also not the power in the land it used to be. Power has been devolved to three of the four constituent nations of the United Kingdom—Northern Ireland (when there is political agreement), Scotland and Wales—to such an extent that London's room for strategic manoeuvre has become limited, even if foreign, security and defence policies remain reserved competences. The Scottish government has even used British taxpayers' money to set up shadow "embassies" in foreign countries, and London has allowed it to do so.

London's creeping paralysis is now further complicated by the culture wars imported from the United States and the evident loss of confidence by the "authorities." For example, the police have sufficient powers to prevent the likes of Extinction Rebellion and Just Stop Oil from disrupting thousands of citizens going about their lawful business, and yet police chiefs routinely choose not to apply those laws. Regular pro-Palestinian marches in London have included a small minority inciting race hate against the Jewish community, but the authorities have done little to stop it. Consequently, few citizens in Britain know any longer where the competence of the state starts and ends, and ever fewer of them believe the state is competent at all given the culture of timidity and caution at the top of government, which is the very antithesis of active leadership.

Theresa May's one-time guru, Nick Timothy, is clear about the causes of the crisis in London,

> None of this has come about in a fit of spontaneous disorder and chaos. The root causes are political decision-making with little regard for operational reality; legislation that sets lofty goals without realistic plans to achieve them, making ministers vulnerable to judicial second-guessing; macro legal frameworks that politicians understand cause real problems yet lack the intellectual courage to change; international treaties that are treated as de facto constitutional laws; and the surrender of executive power to an administrative class that takes political decisions—police chiefs, quangos, regulators, and supposedly expert committees among others.[5]

At the core of this crisis is a fracture in government over the purpose of power: leadership or management. Sir Lawrence Freedman captures this confusion neatly, "Instead of the deliberate decisions of a few, critics [*of strategy as a concept*] pointed to the countless moves of innumerable individuals, unable to see the big picture yet coping as well as they can in the circumstances, leading to outcomes that nobody had intended or desired."[6] Yet the stated aim of National Security Strategy 2010 (NSS 2010) was

> to be able to act quickly and effectively to address new and evolving threats to our security. That means having access to the best pos-

> sible advice, and crucially, the right people around the table when decisions are made. It means considering national security issues in the round, recognising that when it comes to national security, foreign and domestic policy are not separate issues, but two halves of one picture.[7]

Earlier, the National Security Strategy 2008 had stated that it would, "set out how we will address and manage this diverse though interconnected set of security challenges and underlying drivers, both immediately and in the longer-term, to safeguard the nation, its citizens, our prosperity and our way of life."[8] But between 2008 and 2010 a rupture took place in the ability of government to afford its strategic security ends, which were not at all reflected in what purported to be national contingency planning documents.

The Virtual Ten-Year Rule has existed since the end of the Cold War. Irrespective of the strategic environment successive governments have preferred to accept more risk for Britain and its people rather than investing in defence. The approach to security has been about the appearance of managing priorities, risks, and threats rather than in fact managing them. Consequently, all strategic security and defence reviews since 2010, in whatever form they have taken, have been devoted to saving money whilst pretending to defend Britain. They have taken little real account of just how quickly things could become dangerous for Britain—the Virtual Ten-Year Rule. Perceived threat, therefore, tends to reflect affordability rather than probability.

When Strategy Meets Reality

What happens when poor strategy meets hard reality? In 2014, Russia seized Crimea from Ukraine, and in July of that year a Russian Army SA-11 anti-aircraft missile shot down a Malaysian Airlines Boeing 777, killing 298 passengers and crew including some British citizens. In 2018, the GRU, Russia's military intelligence service, undertook an attack in Salisbury using a lethal nerve agent, Novichok, which ended up killing two British citizens and threatening the lives of countless others. This attack followed the 2006 poisoning in London of Russian dissident and exile Alexander Litvinenko using Polonium 210. Why? Because Putin regards Britain as weak and poorly led. The attacks also revealed the extent to which London has repeatedly got its security and defence assessments dangerously wrong in that what was deemed least likely to happen did happen but had not been prepared for. The tragedy is that despite all the evidence a thorough trawl of all recent security and risk assessments reveals that little has changed, and London is still not taking the threat of direct Russian aggression seriously enough, most notably the very real threat Russia poses to Britain's vulnerable undersea energy and communications infrastructure.

All national security strategies are trade-offs between affordability and capability but for too long the way London has defined affordability has broken the link between threat, strategy, security and defence, and between capability and capacity. Such constraints have also led to a profound failure of imagination that is all too evident in Britain's recent national security strategies. The sheer pace and nature of change requires a much more radical understanding of the strategic environment if adverse strategic change is to be resisted and shaped. In that sense, Integrated Review 2021 and Integrated Review Refresh 2023 are "worthy" successors to NSS 2010 and NSS 2015 because whilst they highlight global trends and challenges, they offer at best partial solutions. They tinker with the international system at a time of profound geopolitical tension. One of the many paradoxes of the latest reviews is that they all use the Russo-Ukraine War to claim a return to a resource-led rather than a threat-led approach to security and defence funding, even though that clearly is not the case. Therefore, London fails to address in any meaningful way the vital relationship between Britain's security, British national means, and Britain's necessarily large strategic ends, and produces essentially political rather than strategic documents.

All British strategic reviews of the last fifteen years or so have also tended to offer a broad sweep of the security environment and emphasise the non-military nature of many of the challenges and threats. They have also revealed a profound tension between Britain's European political instinct to civilianise security policy as much as possible, and an American tendency to militarise security policy as early as possible. To square that unsquarable circle, no "strategy" has offered a clear set of costed planning priorities for dealing with high-end military threat, just aspirations. This lacuna is not only reflective of Britain's lack of strategic autonomy from the US over a long period, and the loss amongst the elite of what might be called a strategic culture, but also the struggle in government between Atlanticists and Europeanists.

Since 2010 London has repeatedly failed to give proper guidance to security planners beyond very general guidance and hectoring them to find eternal savings and efficiencies (cuts). Rather, London's strategies only operate at the declaratory level of policy with the focus on making government look good and protecting ministers from criticism. This conflation of the public good with political need has driven endemic short-termism across London and a belief that defence expenditure is forever discretionary. Comprehensive spending reviews are too often little more than a government euphemism for real-time defence cuts with the main political aim being to prevent any government being tied to any particular security commitment beyond the most informal. NATO is simply taken as a given, whilst other "strategies," such as AUKUS (the trilateral security partnership entered into with Australia and the US in 2021), have more to do with opportunism than settled strategy.

The entire national security strategy process implies that not only defence but also Britain's wider security is discretionary and not a priority and ranks markedly lower than other demands on the national exchequer, such as health, education, and welfare. This dangerous bottom-up approach to the relationship between British strategy and British security and defence is driven by metrics of affordability that have little to with reality—although this attitude is perhaps less pronounced when it comes to investments in cyber-defence, the security fashion of the moment, and a threat still ill-defined. The inherent tendency of London to place the financial cart before the strategic horse is nothing new. The very question of what comes first, money or strategy, is *the* chicken-or-egg question all states must confront (the answer is both, for they are two sides of the same coin). Successive British governments have simply chosen austerity and the here-and-now in place of strategy and what is dangerous, when in fact austerity requires more strategy not less. It is as though austerity has become *the* alibi for London's penchant for declinism and pessimism, particularly in the wake of the COVID crisis.

That is why national security strategies and their ilk are less drivers of planning and more exercises in public diplomacy and strategic communications. Whatever specific proposals they make are too often an unhappy settlement between the main spending departments (welfare, health and education) and those departments of state charged with external engagement (the Foreign, Commonwealth and Development Office and Defence). The problem with such domestically driven political trade-offs is that it is at precisely such moments of economic recession that threats to security grow. For many years Britain's national security approach has thus been a decidedly inside-out, bottom-up approach to strategy that too often offers little or no basis for proper whole-of-government planning and thus no ability to achieve a reasonable balance between the ends, ways and means of national security and defence. Such a parsimonious approach to big issues is also reflected in all the recent security and defence strategies which are again a curious mix of partial analysis, unclear and ill-resourced intent, political rather than strategic aspiration, and spin.

The responsibility for such failure ultimately rests with political leaders who repeatedly refuse to allow those charged with drafting the strategies to address the only question that really matters: what are Britain's strategic security interests and how can they be realised? Muddling through is fine for a country at the very top of power, but not a particularly good idea for a state that is no longer central to the calculation of allies and partners and is no longer taken sufficiently seriously by potential adversaries to stop them exploiting and attacking Britain's many vulnerabilities.

Never Learning Lessons

The Integrated Review Refresh 2023 (IRR 2023) says that Britain will "continue to draw lessons on an ongoing basis, including from Ukraine, and this will be essential to future plans for the armed forces, particularly in the land domain."[9] In that case, why does the Defence Command Paper 2023 cut the army to its smallest size since before the Napoleonic Wars?

Britain is in the worst of all strategic worlds: responsibility without capability. Britain has just enough power and influence to be a target, but insufficient power, history, and influence to deal with the many threats it faces. All of Britain's latest strategy documents, most notably the IRR 2023 and Defence Command Paper 2023, suffer from essentially the same faults, which is hardly surprising given their provenance. They all lack a sense of urgency, place too much emphasis on the wrong threats, are driven too much by the politics du jour (the green agenda and climate change), exaggerate Britain's strengths, assume the

Americans will always bear much of the British security load, and continue to treat risk and threats through bureaucratic stovepipes and fiefdoms of ministries. They also miss the biggest threat of all: the sudden impact of several threats simultaneously across the conflict spectrum, which, given the fragility of contemporary Britain, would rapidly lead to paralysis and potential destruction.

The hard truth is that London is unable or unwilling to adapt to the very change it describes, far too focused on the short-term, and resistant to any action that might offend the many activists and special interest groups that now infest London. In other words, Britain is weak when and where its government pretends to be strong. London is willing to send a ship to the Gulf or half a brigade to Kosovo but only under the umbrella of American protection. For reasons of self-grandeur and history Britain has become a weak state for which strategic pretence is underpinned by defence pretence. What happens if the umbrella collapses?

The latest strategy documents are also strong on describing what threatens Britain, but deeply unambitious about the ends, ways, and means needed to meet such threats. Indeed, if all national security strategies since 2010 were to be treated as trend analysis, then the message is clear: the nature, scope, and pace of change and threat is changing far faster than Britain's ability to deal with them. Worse, political and bureaucratic inertia is evident in all the recent reviews, with the relationship between prescription and outcome too often little more than short-term politics masquerading as strategy.

IRR 2023, for example, is little more than a creatively stitched together audit of everything ministries are already doing to create the impression of a coherent strategic narrative, the coherence of which is not always apparent. It presents itself as "building on" "Global Britain in a Competitive Age: The 2021 Integrated Review of Security, Defence, Development and Foreign Policy."

In fact, its prescriptions, far from being an update, simply reveal the weakness of the assumptions that underpinned the original review and the nature of the bureaucratic politics to which all such strategies are subject, as well as the extent to which HM Treasury sees security and defence as a cost rather than a value. The "Updated Strategic Framework" in IRR 2023 is more heat than light, proposing a profusion of strategies with few resources to realise them. "Narrative" is more important to them than delivery. In addition to the Defence Command Paper 2023, there is a Defence and Security Industrial Strategy, a National Artificial Intelligence Strategy, a National Cyber Strategy, a National Space Strategy, a Net Zero Strategy, and an Arctic Policy Framework.

With so many priorities, nothing is a priority. This profusion of priorities is primarily due to the sheer scope of political activities implicit in the "Global Britain" mantra. The Euro-Atlantic is a priority, but so is Britain's role in the Indo-Pacific, and Britain's future role in the Comprehensive and Progressive Agreement for Trans-Pacific Partnership or CPTPP, over which Britain will have little or no influence. The defence-strategic concept at the heart of IRR 2023 is entitled, "Deter, Defend and Compete Across All Domains," but contemporary deterrence and defence domains require a capability and capacity to generate effect across air, sea, land, space, cyber, information and knowledge, each of which is hinted at but with little specification and no prioritisation. The central weakness of the review is that it does not address the essential security dilemma Britain faces: how to resolve an inability to both project power *and* protect people in a way relevant to the very threats it describes. This lack of intrinsic relationship between power projection and policy is most evident in the section, "Addressing Vulnerability Through Resilience," with little evidence of the restructuring of governance that delivering such ambition would require. The review is similarly uncertain with respect to generating strategic advantage: where, to what end, and with what?

In short, London's big 2023 integrated review of security, defence, development, and foreign policy is solid analysis presented very glossily, but it is a wish-list, full of contradictions and at times plain vacuity. It is an apple pie and motherhood "strategy," written first and foremost for political effect and, as such, an exercise in government strategic communications at best, "spin" and national self-delusion at worse. It contains little or nothing of substance about what is really needed to protect people in the face of emerging, emergent, disruptive, and destructive threats. What power projection is implied suggests soft power is as vital as hard power. There is also much reference to integration of effort, even if there are few of the levers London would need for such a whole-of-government approach to work.

Defence strategy is a subset of security strategy and because IRR 2023 is so weak on the fundamentals so is the Defence Command Paper 2023, particularly where it addresses the impact of the cost of the nuclear deterrent on Britain's conventional defences. Both papers continue to insist the cost of the deterrent must be funded from the defence budget. And yet there is also an urgent need to fund an increasingly technology-hungry army, navy, and air force. It is hardly surprising that defence cost inflation is soaring and the unit cost of both platforms and systems with it. Something had to give. It did. Since 2010 London has abandoned quantity (which in warfare has a quality in and of itself) for some ill-defined concept of "digital" quality. And yet the lessons from the war in Ukraine scream out for far more of both quality *and* quantity. London even agrees when the IRR 2023 states that, "The risk of escalation is greater than at any time in decades,"[10] even if transnational security challenges (terrorism) still pose a considerable risk to Britain.

In other words, to close any inconvenient gaps between the scale of force and the fighting power it must generate to be credible in its assigned role, London has retreated into an old fantasy:

that science can offset a lack of force. Science and technology are indeed transforming the character of warfare, but the suggestion that new wonder weapons can replace traditional but still vital capabilities, such as artillery, engineering, mass, and logistics is profoundly wrong. And yet, this is precisely what many at the most senior levels of the Ministry of Defence, on both the political and military sides, are claiming. They need to ask themselves a question: is defending Britain or pleasing their political bosses more important?

How Much Threat Can Britain Afford?

HM Treasury has consistently made resource rather than threat the main driver of planning defence policy and strategy. So, how much threat can Britain afford? London has consistently downplayed the extent to which big, long-term security and defence threats need big, long-term security investments, deeming such threats as least likely for political convenience, even as the evidence suggests otherwise. All of London's recent strategies have failed the very test of strategy: to meet the threat of observable reality. Rather, London finds itself bogged down in the mud of incrementalism and partialism.

To prosper in a complicated and contested world Britain at the very least needs "joined-up" government, the absence of which is evident in the lack of consistency and quality in government strategic communications. As Paul Cornish, Claire Yorke and Julian Lindley-French wrote,

> If strategic communications are to be truly national, they must reflect not only government policy and an executive message but a national narrative that is understood, owned, and endorsed across society. Strategic communications must be seen to reach out from central government to operational environments (both military and non-military) and to the local domestic constituency. Equally, they

> must be perceived to be relevant, credible, and authoritative at all levels of the governmental process, from the highest policy level to the practical levels where engagement takes place. Finally, to be effective they should be both a "centre of government" concern (i.e. an organic part of the policy-making and strategic process at the highest levels) and a "whole-of-government" unifier (i.e. a common feature of all activity at all levels of government).[11]

They are not.

Threat perception is driven as much by vested interest as disinterested hard analysis. Since 9/11 London has become packed with advisors and experts on terrorism, organised crime, human trafficking, overseas development, and cyber. Experts on defence and deterrence (and Russia) are few and far between in government. There are also many think-tanks and think-tank reports, but unless their funding comes from government—in which case both the topic, and too often the conclusions, are precisely what government wants to hear—they have little traction. Consequently, London is a hotbed of groupthink where everything is there but, for parochial reasons, not in the right order of priority.

Consequently, Britain's security and defence effort lacks balance and reeks of vested bureaucratic interest, most notably the strength of the counter-terrorism lobby. If London was capable of effective analysis the only possible strategic conclusion it could come to is that Britain needs a new comprehensive civil defence concept and structure overseen by a Ministry of Civil Defence, allied to far more capable conventional armed forces, if future war is to be deterred. This is because whilst no democracy should be in the business of war in the way autocratic regimes are, if they wish to preserve the peace they must be in the business of deterrence, which Britain is not. Much of "civil protection" is left in the hands of the over-mighty but chronically incapable Home Office, whilst national defence is the responsibility of a Ministry of Defence in desperate need of profound reform, and a Treasury full of economists who would rather pay for none of the above.

Furthermore, what are cited as the lowest existential risks to Britain should be the top priorities. The reasons for this inversion are clear. There is profound confusion in London about the relationship between the most likely risk and the most catastrophic threat. Prioritising the latter would require wholesale change which would cost a lot of money at a time of financial constraint (when is it not?). Government would need to admit to having made a series of false assumptions and taken many missteps across many years. A more interest-driven analysis of threat and risk would undoubtedly lead to a radical change of focus and structure which would also mean that many in the London bureaucracy would lose out, not least the very influential counter-terrorism lobby. Any serious analysis of the strategic environment would suggest Britain is entering an era in which defence and deterrence allied to comprehensive civil defence should be the "Order of the Day," but this is given little weight by those charged with Britain's security. At the very least, Britain would need a very much strengthened and restructured National Security Council and National Situation Centre.

Take the National Risk Register (NRR) 2022. Of its sixty-four sections eleven are devoted to "Terrorism," twenty-eight to "Accidents," although they include drone attack and space-based attacks. There are eleven sections devoted to "Natural, and Environmental Hazards," whilst eight sections are devoted to what is called "Humans, Animal and Plant Health." Only one section is devoted to "Geographic and Diplomatic" risks, whilst three sections are devoted to "Societal" risks. Finally, at the very end three sections consider "Conflict and Instability" (sections 61–4 of 64). By placing human welfare, behavioural impacts, impacts on essential services, economic and environmental damage before security impacts, NRR 2022 reveals the mistaken ordering of priorities at the heart of British contemporary national security.

Crucially, the "Updated Strategic Framework" makes no proposals for structural change in government to meet the challenges it cites. Rather, it is assumed the Home Office will be responsible when it is already both overloaded and inefficient, and when what is needed is a Cabinet-level Minister of National Resilience overseeing a Ministry of National Resilience. Even the language employed in the framework reveals a lack of strategic ambition, which is really the true ethos of recent "strategic" documents. What should be a new civil defence concept is simply referred to as "civil protection" whilst there is no over-arching approach for the protection of all critical national infrastructures and people in the event of multiple and simultaneous attacks. The framework claims that "The new Resilience Directorate in the Cabinet Office will drive the implementation of the measures set out in the framework and develop our ongoing resilience programme."[12] And yet, the Cabinet Office has never carried the necessary weight to implement such a directive across the government machine and as Michael Gove admitted during the COVID enquiry there is little evidence that Downing Street is willing to give it such powers.

British national security strategies in whatever form they take are more about political legitimisation of contemporary policy than about British national security. They reflect a culture at the heights and heart of government that is concerned more about the management of decline of a past Britain than preparing this Britain for its rightful place and role in the future world. They are thus more about parochial bureaucratic politics and short-term trade-offs than preparing Britain and its people for the intense strategic competition ahead. Britain's support for Ukraine? Not only is Britain giving the Ukrainians just enough to stay in its war with Russia (and barely that) rather than win, but it is also (its claims to the contrary notwithstanding) using the British armed forces as a cash cow to fund such an effort.

4

ENDS, WAYS, AND HAS-BEENS?

Defence-Strategic Pretence

What happens when defence strategy fails? In early 2024 the disarray caused by a severely over-stretched British defence budget became apparent. In January 2024 it was announced that HMS *Argyle* and HMS *Westminster* would be decommissioned because there were not enough sailors to crew two Type 23 frigates as well as the new Type 26 and Type 31 ships that will be delivered to the Royal Navy over several years into the 2030s. This is despite HMS *Westminster* very recently undergoing a costly and major refit (which was abandoned midway), and comes at least four years before their successors, HMS *Glasgow* and HMS *Cardiff*, join the fleet. Early on 12 January, a Type 45 destroyer, HMS *Diamond*, together with four RAF Typhoons carrying Paveway guided bombs, operating out of RAF Akrotiri in Cyprus, were part of a US-led coalition which launched missile strikes on several sites in Yemen. The Iranian-supported Houthi regime was attacking Western shipping bound for the Suez Canal and passing through one of the world's sea-lane choke points, the Bab al-Mandab.[1]

On the same day, Prime Minister Rishi Sunak, visiting Kyiv, announced a £300 million increase to British military aid to Ukraine. Then, in the same month, news was leaked that two Royal Navy Assault ships, HMS *Albion*, and HMS *Bulwark*, would be withdrawn ten years earlier than planned. First, such a withdrawal would have effectively ended Britain's already limited amphibious capability, even though the short-range F-35Bs only afford Britain's Carrier Strike Group a support role for littoral strike operations. Second, the Foreign Secretary Lord Cameron had just said that the world was more dangerous than at any time for decades, citing precisely the kind of threats for which Royal Marines train. The usual disingenuous rationalisation was offered that the commandos could operate from the two aircraft carriers. In the end, London backed away from the decision but the idea that the Royal Navy was forced to think of such action reveals the stress the force is under. It also revealed the defence pretence at the very top of the Sunak government. A government which clearly saw defence as a cost not a value.

Three further defence-strategic tensions were also highlighted by these episodes. First, the need for Britain to play a proportionate role to protect sea-lanes of communication. Second, the seemingly endless friction between the different elements of the defence establishment over whether Britain should focus on a land- or maritime-based defence strategy and shape the future force accordingly. Third, London's "solution" of funding shortfalls, which cuts capability irrespective of the threats Britain faces and thus builds risk into defence, most notably for those under-equipped servicemen and women who must carry out London's bidding. Britain does not need a land-based or a maritime-based defence strategy per se but rather it needs to align defence strategy with its critical interests to meet critical threats. This means armed forces scaled to meet at least the most likely and serious threats, and of sufficient capability, capacity, and quality to exert influence over vital force multipliers such as the United States, NATO, and democratic partners elsewhere to meet all possible contingencies.

On the face of it, the macro-economics of Britain's defence strategy do not seem inappropriate, especially since London's announcement to spend 2.5% GDP on defence by 2030, but the micro-economics tell a different story. Put simply, Britain these days punches below its geopolitical weight.

Even before the decision to increase the British defence budget to 2.5% by 2030 it was planned to be higher than that of France until at least 2027, Defence Command Paper 2023 can best be summed up thus: ever higher ends, limited ways, but only a few more means. British defence expenditure will also increase from 2.06% of GDP in 2019–20 to 2.27% in 2024–25 (or 2.35% if annual spending in support of Ukraine continues at 2023 levels). However, the core defence budget will remain at 2.2% in real terms until 2025, even though core capital spending between 2019–20 and 2024–25 increased by 62% in real terms. Core resource spending, the measure of the resource available to fund delivery, will fall by 4%, driving the further reduction in personnel, particularly in the army.[2]

The "Defence Nuclear Enterprise" allied to high defence cost inflation is the cause of this imbalance and goes back to then-Chancellor of the Exchequer George Osborne's disastrous 2010 decision to place the cost of the nuclear deterrent within the conventional defence budget in the wake of the 2008 banking crisis. The Defence Nuclear Enterprise is also in a mess. Matthew Harries writes, "The UK's facilities for making and maintaining nuclear weapons, wrote [Dominic] Cummings in a recent post, are characterised by 'rotten infrastructure' and 'truly horrific bills' amounting to 'many tens of billions' over the coming years."[3]

Perhaps Osborne could have spread the cost by adding it to the bill the banks should pay for the public largesse from which they benefitted following the collapse of "casino banking." Of the

£11 billion extra money allocated to defence over the 2020–25 period £9 billion will be spent on the nuclear deterrent, the nuclear share of the "Ten Year Equipment Budget" rising from 25% to 34%. Money is also being syphoned off from the defence budget under cover of the digital domain, most notably to the National Cyber Security Centre. London's April 2024 decision to increase defence expenditure to 2.5% would mean at least an additional £20 billion (the £75 billion in additional moneys claimed by Sunak is questionable). But as Dan Sabbagh writes in the *Guardian*, "Boris Johnson had previously promised to increase spending to 2.5% by 2030 but, mindful of other spending pressures and the military's long record of financial waste, Sunak has been vague on timing."[4] Former US Major General and NATO Deputy Assistant Secretary General for Defence Investment Gordon Davis also points out that the profound challenges Britain faces are not merely financial and short-term:

> The deterioration has occurred over decades as critical threats were ignored or underestimated in terms of what was needed to deter or defeat them in case of direct aggression. It will take sustained and increased defence investment for many years to come to reverse the deterioration and re-establish credible deterrence and defence.[5]

Worse, in December 2023, the National Audit Office published a report on the defence equipment plan in the 2023 Defence Command Paper (DCP 2023). The report states,

> The Plan is unaffordable, with the MoD estimating that forecast costs exceed the available budget by £16.9 billion (6%). Forecast total costs on 31 March 2023 were £305.5 billion, compared with the budget of £288.6 billion. This is the largest deficit in the Plan since the MoD first published it in 2012 and contrasts with the previous year's Plan, for 2022–2032, when the MoD assessed that costs were £2.6 billion less than the available budget. Forecast costs have risen by £65.7 billion (27%) compared with the previous Plan, outstripping a budget increase of £46.3 billion (19%). Costs to support the

> nuclear deterrent exceed budget by £7.9 billion, and the budget for conventional equipment is £9 billion less than forecast costs.[6]

If the purported year-on-year increase to 2.5% GDP on defence must fix anything it is this shortfall. That said, hard choices need to be made and perhaps now is the time for Britain to make the hardest choice of all. Does it need and can it afford a single-mission, bespoke nuclear deterrent or, given the pace of emerging, disruptive, and destructive technologies, does deterrence itself needs to be reconceived? If Britain does want to retain a sea-based nuclear deterrent, would it make more sense to invest in cruise missiles with hypersonic propulsion that would in time enable nuclear-powered attack submarines to carry a nuclear strike capability, as they do today in several navies? After all, the continuous at sea Deterrent is meant to afford London an assured second-strike capability, but with only one submarine at sea at any one time and with the development of AI-enabled sensors, quantum computing, big data and long-range hunting and loitering unmanned underwater vehicles (UUV), it is a bold assumption that the new *Dreadnought*-class submarines can maintain their invulnerability to 2062 when they are due for replacement. Moreover, the *Vanguard*-class is estimated to run an 8% risk of detection on any patrol, but undersea topography also forces deep-diving submersible ship ballistic nuclear submarines to funnel into certain areas. In other words, Britain has an "all eggs in one basket" nuclear capability which is likely to become progressively more vulnerable to discovery and destruction. The hypersonic cruise missile option would build redundancy into the deterrent and enable the new and extremely expensive *Dreadnought*-class submarines to undertake tactical as well as strategic roles, or allow us to simply cancel the programme and invest in more SSN-R (Submersible Ship Nuclear Replacement), the planned post-2035 replacement of the Astute-class nuclear attack submarines. Such an approach would certainly give the submarine fleet far more operational flexibility.

The consequences of an unaffordable defence plan are plain to see in the DCP 2023. The British Army will fall from 76,500 to 72,000 full-time personnel or eight brigades (a British Army brigade is approximately 2,500 personnel) which will be structured into two heavy brigades, a combat aviation brigade, a deep reconnaissance strike brigade, two light brigades, one air manoeuvre brigade, a security force assistance brigade, plus a new Ranger regiment. And yet, under the NATO Defence Plan Britain is required to field two divisions in short order should the Alliance suffer an Article 5 attack.[7] Whither NATO?

At the same time, the risk of major war between the Western democracies and the autocratic states will likely grow well into the 2030s. This threat was made explicit in NATO Strategic Concept 2022, which Britain signed off, and implicit in the Integrated Review Refresh 2023 (IRR 2023). Indeed, it was one of the principal reasons why the review of the review was commissioned (it is never a good idea to let the authors mark their own homework). Given what has happened to Ukraine and what is happening in Russia the military deterrence of Russia (and increasingly China given its impact on the Americans) will remain the principal mission for NATO and Euro-Atlantic security, with Britain, like it or not, to the fore. The worst-case for which Britain should plan would be a simultaneous and engineered full-scale war in both the Indo-Pacific and Euro-Atlantic theatres. Washington is increasingly concerned that without the proper sharing of defence and deterrence burdens their overstretched forces would not be able to cope with such simultaneous high-end engagements, in the process reducing the threshold for the use of nuclear weapons.

Britain's Retreat from Defence Realism

It is not until one considers the recent history of Britain's defence strategy that the extent and scope of London's retreat from realism, let alone strategy, become dangerously clear. In 2017, the House of Commons Defence Select Committee wrote,

> We seriously doubt the MoD's ability to generate the efficiencies required to deliver the equipment plan. In the past, the MoD has proven incapable of doing so—for example, in 2015, when only 65% of planned efficiency savings were achieved. Even if all the efficiencies are realised, there will be little room for manoeuvre, in the absence of sufficient financial "headroom" and contingency funding. This is not an adequate basis for delivering major projects at the heart of the UK's defence capability.

Such weakness was apparent long before 2024 and has been implicit in all strategies from the 1998 Strategic Defence Review to IRR 2023 and the DCP 2023. Why? Threats and tasks assigned to the armed forces have grown far faster than either capabilities or capacities, especially in the wake of 9/11 and the campaigns in Afghanistan and Iraq. Not only have defence planning assumptions remained stuck in a "peacetime establishment" when the country was in effect at war, Downing Street and the Treasury have always sought security and defence on the cheap. Consequently, successive governments have mortgaged the future requirement to pay for the current requirement while there was little relationship between any planned future force and the money needed to afford it.

The management consultancy McKinsey even designed a system which became a byword for ill-conceived attempts to pay for the then-present with disastrous consequences for the now-future—the Smart Procurement Initiative. The Treasury also imposed a government-wide system for the launch and development of projects, all of which built delay and cost into an already strained system. The armed forces are still paying the price to this day.

A radical shift is needed in Britain's defence planning assumptions. First, whilst the United States will remain Britain's closest security and defence ally, the Americans might, *in extremis*, no longer be able to defend Britain and Europe conventionally at one and the same time if they are hard-pressed in the Indo-Pacific and elsewhere. To counter this threat the British would have to do far more for their own defence and that of Europe through a more European NATO. This is not because the Americans are withdrawing from Europe but rather because of the worsening global over-stretch from which US armed forces suffer.

Second, Britain will need to rely more on European allies through NATO and, at the very least, demonstrate to the Americans that Britain is seriously committed to equitable transatlantic burden-sharing. For example, Britain could seek to co-pioneer with France and Germany a high-end, twenty-first-century, "heavy" and fast first-responder European force able to deter and defend in and around the European theatre and across multiple domains.

Third, only Britain and France retain any globally relevant defence-strategic weight in Europe, and it would be logical for Britain to join with France in promoting greater European defence-strategic responsibility, partly by buying into autonomous European strategic enablers. Much will depend on the strength of the Franco-British alliance almost fifteen years on from the 2010 Lancaster House Agreement, which David Richards attended and where he and his French counterpart formally proposed the creation of the Franco-British Combined Joint Expeditionary Force. However, for the halcyon days of the Lancaster House Agreement to return, with the mutual trust required, London and Paris need to put Brexit and its poisonous aftermath behind it. Put simply, neither Paris nor London could secure a good defence relationship without a good political and trade relationship. This will require statesmen and women of stature on both sides of the Channel and high-quality statecraft. One should not perhaps hold one's breath.

Fourth, Britain will face both peer strategic competitors and sophisticated non-state actors employing complex strategic coercion against Britain and its people who increasingly use emerging and disruptive technologies, such as artificial intelligence and a host of other such tech. Therefore, Britain must balance both credible traditional defence and deterrence with effective complex engagement. The mosaic of new threats in an information-digital age will range across 5Ds of disinformation, deception, destabilisation, disruption, and implied and actual destruction and will be little short of warfare in peacetime. Confronting, scaling, and adapting force and resource to meet such threats will also require a much more nuanced understanding of the relationship between civilian-led security and military-led defence, as well as a far more profound, efficient, and intimate partnership between them.

Fifth, for a power such as Britain defence has at least two strategic roles to play: the maintenance of defence and deterrence as a public good *per se*, and the use of defence as a lever of influence over allies. Britain might be able to generate for short periods the high-end capabilities to undertake such roles, but it is unlikely to ever have the capacity to sustain them over time and distance on current budgets. Such tension could well be further exacerbated given the balance Britain might well have to strike between deterring or fighting a short, high-intensity conflict (possibly in mega-urban environments) and engaging in a long, low-intensity conflict.

Sixth, given London's strategic aims, credible deterrence will demand of Britain's armed forces the proven capacity to operate at the high-end and simultaneously across the multi-domain battlespace of air, sea, land, cyber and space, especially if they are to maintain all-important interoperability with US forces when they are under duress. British forces will also need to far better exploit information and knowledge to strategic advantage. This presupposes not only a deeply joint or integrated force (something the UK Strategic Command at least implies), but also a British ability to lead or act as a framework power for combined forces. This might be done either in a NATO context, through perhaps modifying the headquarters of the Allied Rapid Reaction Corps (HQ ARRC) or using it as a model for the HQ of the new Allied Reaction Force (ARF). Specifically, Britain will need the command resources and structures to enable it to act as an alternative command hub if the Americans are busy elsewhere. The challenge for NATO will be to ensure the ARF is more than simply a rebadged enhanced NATO Response Force.

All of the above would require structural changes in London. For example, such a combined (i.e., multinational) joint-force headquarters could only be fashioned by the British if the Service Chiefs of the Navy, Army, and RAF speak with one voice to ministers. Alternatively, the Chief of Defence Staff could assume the authority of Commander UK Armed Forces, in the process relegating the individual service chiefs to an advisory role. This was something David Richards advocated during the Levene Reform process of 2011. Fearing the appointee would be too powerful, the proposal was deemed too radical by ministers and bureaucrats. The pragmatic and experienced single-service chiefs of the day were, perhaps surprisingly, less hostile.

Seventh, technology will drive defence strategy to an unprecedented extent over the next defence planning cycle. Defence futurists tend to exaggerate the speed with which new technologies are entering the battlespace. For example, whilst supercomputing is playing an ever more influential role in warfighting it will likely be a decade and more before the kind of quantum-computing able to drive artificially intelligent swarms of drones is realised. However, such technologies (and many more) are coming and must be factored in, alongside the increasingly "kinetic" impact of offensive cyber. Hypersonic weaponry is already a fact.

Corrosion and Erosion

Over time defence has been repeatedly cut in real terms partly due to delays and cost increases to increasingly complicated systems and platforms. Indeed, equipment inflation is a core component of the defence cost inflation which has proved so corrosive and so eroded the British armed forces. Such erosion has been compounded by the voracious appetite for people, equipment, and money of the major campaigns in which Britain has engaged since 2001, from which the British armed forces have never recovered. There is an ever-growing tension between Britain's defence strategic ends, ways and means with the almost inevitable outcome being another military disaster. This is why British defence strategy, such as it is, lurches from crisis to crisis as the mission and tasks assigned to it continue to grow, but the force and resource available to meet them inexorably and relatively shrinks as the tension between capability, capacity and affordability becomes ever greater. It is not all bad news. For all the challenges Britain's armed forces face they are also probably the best placed of all European militaries to cope with the profound digital change coming their way. Sadly, this more reflects the appalling state of defence in most Western European countries than suggests the British are paragons of defence strategic virtue.

Given the centrality of the partnership with the United States in the UK strategies the very least Britain needs to demonstrate to the Americans is that it has the political will to generate what Washington (not London) would regard as the necessary minimum conventional military capability, which is not at all the case today. Quite the reverse: by withdrawing the army from the continent save for a token presence, by repeatedly cutting the size and capability of Britain's conventional forces, and by spending so much on a new nuclear deterrent force, the hard message to the

Americans and other NATO allies is that Britain is retreating into itself.

Refresh or Re-hash?

This retreat begs a further question: what should the world's fifth or sixth largest economy spend on defence and what kind of force does it need given its location, the threat array it faces, and the alliances and partnerships it needs to leverage?

When launching IRR 2023 then Prime Minister Rishi Sunak described Russia as the greatest regional threat to Britain's security, a threat that was intrinsically linked to the outcome of the war in Ukraine. He also said that China "poses an epoch-defining challenge." Given the scale of the challenges implicit in those statements both IRR 2023 and the DCP 2023 can only really be seen as yet another down-payment on defence, a down-payment on future war, a down-payment on future alliances, and a down-payment on the warfighting lessons from Ukraine, for which most of the British armed forces lack the critical capability and capacity. Is hire-purchase defence, a kind of new Lend-Lease.

Compare IRR 2023 with President Xi Jinping's March 2023 "Great Wall of Steel" speech. "We must fully promote the modernisation of national defence and the armed forces and build the peoples' armed forces into a Great Wall of Steel that effectively safeguards national sovereignty, security and development interests."[8] Xi also announced a further 7.2% increase in the Chinese defence budget to $224 billion (in fact China already spends far more), whilst in 2023 there was a year-on-year increase of some $15 billion.

London would argue that given China's official defence budget represents only 1.5% of its GDP Britain's defence increase is a reasonable and proportionate response. That is not the case because smaller and weaker powers must spend more of their wealth defending themselves. One real problem is the role of economists at the heart of government in London who mistakenly believe that neither China nor Russia really poses an existential threat to either Britain or its allies. Unfortunately, economists by and large fail to understand why wars start. China is clearly gearing up militarily to confront the United States in the Indo-Pacific with profound implications for Europe and its security. Again, given the centrality of Washington to London's security and defence planning assumptions, anything that impacts the US impacts Britain and NATO. One must only read the US National Defense Strategy 2022 to understand the scale of the impact China is already having on American defence planning assumptions.

Ukraine

What is Britain trying to achieve in Ukraine? It is not clear because the military means provided, whilst vital in preventing Ukraine being over-run in the first instance, are nothing like enough to help Ukrainian forces expel Russian forces from their country. Given that reality, sound British strategy should be to assist the Ukrainians in a transition to strategic defence, whilst Britain and her allies develop a truly grand strategic approach to the war by playing a long game that reinforces NATO whilst seeking to divide China from Russia.

The military lessons for Britain emerging from the Ukraine war are increasingly clear and need to be acted upon. First, modern war is a giant black hole into which people and materiel vanish at an alarming rate far beyond that envisaged by the peacetime establishment. At the very least, British forces need far more robust logistics, will need to be far more forward deployed, with enhanced and far more secure military supply chains. Far more materiel is also needed, most notably ammunition, not least because of the rate at which Ukraine has been using up British weapons stocks.

Second, Britain's home base is vulnerable to drone attack. In September 2023, Dan Sabbagh of the *Guardian* wrote,

> Britain's most senior military officer has said the war in Ukraine had revealed a potential vulnerability of the UK to missile and drone attack, arguing it is time for greater discussion about improving homeland security. Admiral Sir Tony Radakin told a defence industry conference in London that there was "an aggressive world out there in terms of state threats," noting specifically that it was becoming easy to "get close to a country and fly drones in." The chief of the defence staff said Britain's armed forces were "having a bigger conversation about homeland defences" and asked whether within that "we need to have a conversation about integrated missile defence. The war in Ukraine has revealed the vulnerability of civilians to deadly missile and drone attack, with Russia relentlessly targeting cities and infrastructure as part of a total-war approach to the fighting."[9]

Third, deployed British forces will also need much-improved force protection. A much-reduced digital footprint—and thus reduced detectability of force concentrations and headquarters ("bright butterflies")—is particularly pressing. The war has also revealed yet again the vulnerability of armour unsupported by infantry and helicopters in the battlespace, as well as the need for British forces to be able to dominate both fires and counter-fires, not only through massed conventional artillery but also by using large numbers of expendable drones, strike drones and loitering systems allied to extremely expensive precision-guided munitions, such as Storm Shadow. Enhanced land-based, protected battlefield mobility is also needed together with increased force command resilience given how often the Ukrainians have been able to detect and "kill" Russian forward (and less forward) deployed headquarters.

Britain's armed forces disappointingly appear to have reverted to historic type since the halcyon days of the 1999 Joint Rapid Reaction Force (JRRF), becoming once again far too "single service," and not sufficiently joint or strategic in outlook, with command too concerned with process as opposed to generating effect. Personnel, both in the armed forces and the civil service, must change the dominant organisation-centric culture and develop far greater agility and flexibility in crises. There is also too much groupthink apparent at the top of government, which prevents sound analysis, and too little strategic education in all its forms and at all levels of civilian and military leadership. Not only does this failing profoundly undermine effective strategy, it also fails to mentally equip leaders at all levels of mission command to understand just why they are doing what they are being asked to do.

David Richards was the principal author of the JRRF concept, when he was working directly for the Service Chiefs and the Chief of the Defence Staff Charles Guthrie. Thanks to his conceptual work together with much learnt on operations and several exercises, he became probably the most experienced joint operations level commander Britain has had in the post-World War Two era. The JRRF work also gave Richards a lot of influence with the single services, all of whom wanted the maximum number of assets in the JRRF and the best possible representation in the Joint Force Headquarters (JFHQ). Readiness was the critical JRRF arbiter. For example, US commanders could not believe how quickly and efficiently the British deployed on Operation Palliser. The same efficiency of deployment was apparent in East Timor, which saw British forces up and running five days before the first American troops arrived from Pacific Command (PACOM) under their poor equivalent of JFHQ.

Service chiefs can no longer simply talk "joint," they need to start walking it again. Doing so would reinforce the fact that there is no need for the profusion of so-called strategic headquarters for a force as small as that of Britain and certainly no need for a Strategic Command. That is the job of the Ministry of Defence (MoD). Rather, the MoD needs to be reforged as Britain's military-strategic headquarters and not the strange kind of politicised super-operational headquarters it has become, with the execution of strategy left to the Permanent Joint Headquarters (PJHQ) and its NATO equivalents. This worked well during the JRRF era and Britain's armed forces should return to simple, operationally proven, and doctrinally pure command arrangements.

Big Cause, Little Effect

DCP 2023 is at best a holding document, which is precisely why it was released in July 2023 as the nation went on holiday. For all the much-needed investment to help defence recover from the damage done to it by several short-sighted governments and two major campaigns, by 2025 Britain's defence peacetime establishment will not only be the smallest it has been for centuries, but in relative terms, it will also have the least fighting power. Indeed, what the strategies reveal is the continuing low importance London really affords defence of the realm, despite their rhetoric to the contrary.[10]

In 2023, the Haythornthwaite Review considered how the British armed forces are to secure the talent they need going forward and to better incentivise those already employed.[11] These issues are extremely important. As Haythornthwaite stated,

> The world has continued to change even in the fifteen months I have worked on this review. The evident problems in recruiting and retaining people with high-demand skills have worsened. Developments in Ukraine have demonstrated how people—responding with agility and flexibility—are a critical part of military capability, even

> in an increasingly technical future. Without action to build agency and agility, I am not entirely confident our Armed Forces will be able to respond when similarly tested.

He adds that "the Ukrainian Armed Forces have amply demonstrated the power of putting large numbers of motivated, incentivised people in the field." The result is that as the risk assessment expands and the military task list grows Britain has ever fewer assets and increasingly disincentivised people to put in the field. It is all well and good speaking of "Integrated Operating Concepts" but for a long time to come technology will enable humans, not replace them (hopefully it never will do given the likely impact on the character of war). The answer is to put strategy before people in that once a strategy is established then and only then is it possible to get the right people.

London also wants to ensure Britain's cutting-edge science, technology and industrial capabilities can be harnessed for effect in the future battlespace. Britain undoubtedly has pockets of such capability but much of the defence industrial base has been lost since 1990. DCP 2023 also aims to increase the productivity of the Ministry of Defence by turning it into a "campaigning department," precisely what it should not be and seeks to rapidly learn the lessons from the Russo-Ukraine War under the heading the "Future is here." However, it is still not clear what lessons from the war in Ukraine will define future war when artificial intelligence, machine-learning, Big Data, synthetic biology, and eventually quantum computing are harnessed and fused as fighting power beyond 2035.[12]

London also highlights the need for what it calls "spiral development," by which equipment is fielded much sooner than planned and upgraded in service. In other words, it places the importance of upgradability at the centre of the equipment concept and thus favours systems rather than platforms. Given the likely entrance of artificial intelligence, quantum computing and machine learning into the battlespace within the life cycle of current and planned equipment, to have any chance of realising such a vision government is going to have to establish partnerships with industry far earlier in the planning cycle than hitherto it must also accept far more of the risk as well as bear much more of the development and production costs. The pressure for government to invest more up front will grow as equipment becomes increasingly complex, autonomous, and unmanned. Perhaps the most ambitious goal of DCP 2023 is the aim to field equipment within five years from initiation, particularly systems (which are developed far faster than platforms). In fact, in the digital domain five years is an eternity and adaptations take place on both the offensive and defensive sides far more quickly on operations.

The three most important words in Integrated Review (IR) 2021 were "competition," "compete" and "agility." IR 2021 was in effect a ten-year plan designed to function much like a US National Security Strategy and had four chapters: science and technology; the open international order of the future; security and defence; and building resilience home and overseas. Its essential purpose was to strengthen the British home base so that London's projection of power and influence becomes more credible, and to adapt the armed forces to reinforce defence, deterrence, and Britain's strategic influence. Some of the enhanced security and defence effects the review also sought involved a fusion of civil and military systems, structures, and technologies.

However, for all the talk of innovation the strategic roots of all the recent reviews were still deep in traditional British statecraft. And for all the narrative about Britain being a "force for good" and a "soft superpower" the review in many ways reaffirmed post-Brexit Britain as a global free trader, mercantilist power armed with nukes. The focus was on enhancing the power of democracies through global multilateralism built on a genuinely grand-strategic effort and investing to that end Britain's still considerable and yet limited resources in security, defence, development, and foreign policy. IR 2021 thus became a kind of "super-Harmel" strategy by both seeking dialogue with regional adversary Russia and trade with global competitor China, whilst also preparing to defend against the very considerable "sub-threshold" threats they pose, as well as the threats of climate change, violent extremism and chemical, biological, radiological and nuclear materials (CBRN).[13] However, by focusing on agility at the higher end of the conflict spectrum Britain is also seeking to reinforce its importance to NATO and the US through a greater capacity to share transatlantic burdens, even as it effectively withdraws from the land defence of Europe. In other words, the politics of these reviews need to be as agile as the force it seeks to create. Given the stated ambition, the ends, ways and means of IR 2021, IRR 2023 and DCP 2023 simply do not add up.

Strategic and domestic political assumptions are apparent in all three reviews and can be thus summarised. First, Britain must in future meet the high-end "force-on-force" challenge across a mosaic of hybrid war, cyber war and hyperwar (ultra-fast warfare that combines a myriad of systems to wreak havoc in an instant) that China and Russia are mounting, whilst also facing a hybrid and cyber war challenge from lesser powers and terrorists. Second, COVID and the search for assured energy supplies have both accelerated and intensified dangerous global strategic competition. Third, the nature and scope of new military technology and the 5Ds of "sub-threshold" continuous warfare (deception, disinformation, destabilisation, disruption and coercion through implied or actual destruction) with which Britain must contend demand an entirely new concept of defence and new methods of deterrence. Fourth, post-Brexit Britain must pivot away from Europe and "tilt towards" the high-growth Indo-Pacific (there is to be a new British ambassador to the Association of Southeast

Asian Nations, ASEAN). Fifth, Britain must invest significant energy to deal with a host of wider transnational challenges, such as climate change, global health, terrorism and organised crime. Sixth, it must use the ship-building programme for the Royal Navy to support jobs in Scotland to weaken the appeal of Scottish independence.

However, between IR 2021 and IRR 2023 a profound political change took place that was the reason for the latter's drafting, even though Downing Street also wanted to mask the extent to which much of IR 2021 was already obsolete. Boris Johnson saw the "cost" as investment, Rishi Sunak did not. Far from the Russo-Ukraine War being the main driver behind IRR 2023, it was in fact the change of leadership in Downing Street. This change is apparent in the defence strategic aims implicit in both reviews. IR 2021 aimed to make the British Future Force the most technologically advanced "agile" force in Europe by 2030 thus affording London a coalition leadership role and post-Brexit influence in Washington, NATO, and other capitals. IRR 2023 were far more modest. IR 2021 sought to demonstrably invest in the defence Special Relationship by creating an integrated force able to operate with US forces at the high end of military effect across air, sea, land, cyber, space (including protection of resilient space-based systems), information and knowledge. This was part of a new concept of deterrent escalation across a spectrum of information war, conventional war, digital/cyber war and "continuous-at-sea" nuclear deterrence to 2070 at least. IRR 2023 accepted the principle but lacks the same urgency.

IR 2021 aimed to generate sufficient information, digital and defence power to enable post-Brexit Britain to exert some continued influence over European and other allies and partners, as well as being a "force for good" in and of itself. IRR 2023 was far more parochial and pragmatic. IR 2021 wanted to generate greater strategic influence through the efficient integration of high-end military force with intelligence, diplomacy, and aid and development to maintain a global presence across the civil-military spectrum, and thus retain a seat as a Permanent Member of the United Nations Security Council. Again, IRR 2023 was more modest. IR 2021 also sought to deliver an increased British capacity to compete below the threshold of armed conflict, to better defend sea-lines of communications (both surface and sub-surface) that are critical to British economic interests, and to restore Britain's defence outreach by rebuilding the network of defence attachés and other elements of defence diplomacy. IRR 2023 was far more about the near and dear and the here and now.

Consequently, Britain is (again) clearly taking a significant risk with its defence-strategic choices in DCP 2023. The implicit post-Brexit message to the Americans is still that alone amongst Europeans it is Britain that will maintain high-end interoperability with the US future force, but at what cost? The message to other Europeans is that even though Britain is a nuclear-armed island the British are still willing to defend Europe (through what it calls "collective action and co-creation") but only if Britain is treated equitably by the EU and only on British terms. Indeed, both IR 2021 and IRR 2023 effectively marked the end of the October 1954 commitment to station a large British land force on the European continent in support of NATO collective defence. There was also an implicit assumption that Britain will never again engage in another Afghanistan or Iraq and is thus pivoting away from land-centric extended stabilisation and reconstruction campaigns. Rather, Britain's commitment to the future defence of Europe will be made through a profound shift of posture to one of engagement across the hybrid, cyber, and hyperwar domains.

The risk? In 2017, when NATO assigned capability targets, Britain committed to the provision of two land divisions. That is now impossible. Whilst it survives the British-led deployable headquarters, the HQ Allied Rapid Reaction Force (HQ ARRC) was by-and-large absent from both reviews, which is strange given the centrality of NATO and the regional threat posed by Russia. A "manoeuvre division" (3 UK Div) will be maintained in the army's order of battle (ORBAT), although it would not be a formation that either US General Cavoli or Russian General Gerasimov would recognise as such. The army will also create four new US Ranger-type battalions like the US Green Berets, which is ironic given that the US Rangers emerged from the Royal Marine Commandos.

All the above points to the transformation of Britain's armed forces into a small, high-end strategic raider force capable of commanding small complex coalitions during limited-span first-responder defence and deterrence operations. However, the British will be unable to conduct extended campaigns over space, time, and contact, due to a significant further reduction in large-scale fighting power. For example, there will be no armoured infantry fighting vehicle, with the Warrior tracked armoured vehicle to be replaced with a lightly armed, battlefield-mobile, wheeled Boxer vehicle. So what presence Britain does maintain in the rest of Europe will be penny-packet formations such as the battlegroup in Estonia that leads NATO's enhanced forward presence (eFP) therein, a reconnaissance squadron in Poland as part of the US-led eFP, and the UK will also continue to support the training of Ukrainian forces, mainly through the Royal Navy.

The Plan

Such a force vision sits uncomfortably alongside the main reason given by London for DCP 2023—the Russo-Ukraine War. It is also noticeable that there is no mention of "Global Britain" or China in DCP 2023. It also focuses on the scope and pace of change rather than the nature of the security environment.

Above all, DCP 2023 considers the lessons from the Ukraine War and the consequent revised relationship with industry if Britain is to realise the stated force goals. Herein lies the problem: then-Defence Secretary Ben Wallace needed a further £11 billion from the Treasury to cope with the consequences for the British armed forces of London's extensive materiel support for Ukraine. He got £5 billion, and it is interesting to note that by February 2024, two years on from Russia's invasion of Ukraine, London had donated some £4.6 billion of military aid to the Ukrainians that might otherwise have been spent on modernising Britain's underfunded military. The impact of the planned post-April 2024 additional £20 billion for the military budget will depend on when it is invested and on what.

There is a wider question that whilst uncomfortable must also be addressed. Is arming Ukraine with very significant numbers of British weapons correct when Britain has little or no defence industrial surge capacity to replace them? It is a tough question but given the world into which Britain is moving, and recognising the scale of the tragedy Ukraine is facing, what of British critical interests? The planned £2.5 billion to be spent on replenishing munition stocks depleted by transfers to Ukraine is to be welcomed because it reinforces the £560 million committed in 2022. London is, in effect, making a down-payment on the lessons that must be learnt from the Russo-Ukraine War, but given the extent, scope and cost of those lessons it is not much of a down-payment.

This is because really learning the lessons of the war in Ukraine will cost the British a lot of money and demand London shift its emphasis from maritime-air-digital back to land and transform the British Army from a counterinsurgency policing force to a full-on warfighting force. Britain will now invest in "munitions infrastructure" to accelerate the acquisition of ammunition, which has been consumed at a far higher rate in Ukraine than expected. However, nothing short of a revolution will be needed in the British defence, technological and industrial base far beyond the changes envisioned in the 2021 Defence and Security Industrial Strategy.

Britain does not so much need a defence industrial strategy as a defence industrial revolution if the ends, ways and means implicit in the reviews are to be realised. In November 1933, the Defence Requirements Sub-Committee (DRC) was formed to consider the shortfalls and deficiencies in Britain's armed forces. A new version of the DRC needs to be stood up as a matter of urgency because Britain's defence procurement is also (eternally) in desperate need of root and branch reform.

For the strategies to be realised acquisition cycles will need to be markedly accelerated and unit procurement costs of both platforms and systems markedly reduced, or an awful lot more taxpayers' money must be spent and inevitably wasted. There is also a profound and growing tension between the acquisition of platforms and systems, which tends to take about five-to-seven years, and the fact that technology evolves every five-to-seven months. As the war in Ukraine is demonstrating, European states and even the US simply lack the defence industrial capacity to ramp up production immediately and rapidly.

At the October 2022 Future War and Deterrence Conference, the message was clear:

> A new and far more interactive and proactive partnership is needed between government, defence industries and the wider military supply chain. Such supply chains also need to be made more robust and secure. The pace and scale of political, economic, and military-technical change risks undermining Allied cohesion and force interoperability, as well as keeping security and defence planning in democracies off-balance, with long-term project management a particular lacuna.[14]

British forces need rebuilding, not simply reinforcing, and that will mean a properly designed and managed long-term capital programme. Nothing less will suffice. Any such programme will never be realised without a new and far deeper partnership with industry, which will in turn need contracts that are both longer and more stable than hitherto. This is because both military platforms and the systems that sit on them are about to undergo a technological revolution in which speed of data will drive speed of information, which in turn will dictate both the speed of command and its relevance on the battlefield. Britain is not only going to have to spend more but the British defence technological and industrial base must be extended far beyond existing stakeholders as part of a radically reconceived supply chain. In other words, "Defence" will have to reach out to new tech communities and learn to operate at their tempo.

Procurement will also need to be far more agile than the bespoke prime contractor "cost-plus" method which still endures despite London's claims to the contrary. Smart procurement is not smart if it simply mortgages the future of defence. There also needs to be far more bought "off the shelf," much in the way the Poles have procured tanks and other equipment from South Korea. The 2021 Defence and Security Industrial Strategy emphasises innovation as so many of its predecessors have, and the need for a robust and appropriate skills base. Indeed, because the emphasis is on the development of the future force it also highlights the vital need to exploit the entire national knowledge base and supply chains far beyond the traditional defence-industrial sector.

Much of the focus is on major investment in science and technology, which is expected to reach 2.7% GDP by 2025, with the fusion of national and defence investment strategies the focus across the civilian and military spectrum. Much of the planned military investment will go into new technologies such as space systems, cyber, directed energy weapons, AI, biotech, and quantum computing. Much is also made of the need to build supply-chain resilience by actively defending it, imposing tougher requirements on foreign direct investment, creating a national tech-industrial base, harmonising trade and export practices with the US, and looking to better exploit multilateral institutions and bilateral relationships. Critically, a defence Artificial Intelligence (AI) strategy and a defence AI centre will also be established together with investment in what the paper calls a Defence and Security Accelerator (DASA) to identify innovative solutions to key challenges.

The question then becomes how. There are lessons from the past that point to options London should explore. On 24 February 1936, Prime Minister Stanley Baldwin rose in the House of Commons and said,

> We have, as the House is aware ... a very great problem [German and Japanese rearmament] ... that will have to be met in the next four to five years and, as we go on to meet those conditions, one of our greatest problems will be to consider whether such measures as we have taken hitherto will be sufficient.[15]

That same month Britain began rearming and over the next eight years both modernised the Royal Navy and created the world's most advanced air defence system, which the Luftwaffe discovered to its cost in 1940.

The parallel between 1936 and 2024 is that it was (again) the British Army that by and large lost out. There were several reasons for this, perhaps the most telling of which was a political determination in London never to return to the trenches of World War One in which several of Britain's leaders had fought between 1914 and 1918. And although the British had in effect invented "Blitzkrieg," or the "All Arms Battle" as it was known in 1918, by 1935 the British Army had resorted to being what it had been since 1815—an imperial policing force. Britain also assumed that France would mount the land deterrence and thus defence of Western Europe in the event of war with Germany because it had Europe's largest and, on paper at least, most powerful army. Unfortunately, the French Army was riven to the core with rivalries.

Consequently, in May and June 1940 Lord Gort's British Expeditionary Force (BEF), which represented only 10% of the Anglo-Belgian and overwhelmingly French forces defending Northwest Europe, was forced into a calamitous retreat. When the French fortress of Sedan fell on 15 May 1940, Gort's force was too small and lacked the necessary joint fighting power to act as an independent force and was forced to retreat to Dunkirk where, despite the eventual evacuation of 330,000 men, the British Army lost a lot of men and the bulk of its mechanised equipment. Whatever the quality of the BEF, it was simply too small and too under-equipped to make a qualitative difference on the ground in the face of the Wehrmacht's onslaught. Today?

The reviews all imply Britain must prepare for war, or at the very least a very uncertain and fragile peace, but do so with a determinedly peacetime mindset. In fact, only something akin to the British Shadow Factory Plan will work because it enabled London to give the British peacetime economy the option to move quickly to a war footing and thus quickly increase war production in September 1939. In 1935, the "Shadow Scheme" was established by the British government. The aim was to subsidise manufacturers to construct a system of new "shadow factories" reinforced by additional capabilities at existing aircraft and motor industrial plants that could immediately increase war production on the outbreak of war. It was this scheme that led rapidly to radar, the Hurricane and Spitfire fighters and eventually the Lancaster bomber. It also enabled Britain to surpass Nazi Germany in aircraft production in June 1940, a lead Britain never lost not least because of the entry into the skilled workforce of millions of British women.

Improved efficiency was also as important to Britain in 1940 as it is to Britain today. Government control in the late 1930s was exerted in a way that is simply not the case today and immediately began to find savings and efficiencies. For example, the Ministry of Aircraft Production had an immediate galvanising effect. Upon taking over Royal Air Force storage facilities it was discovered that whilst the RAF had accepted over 1,000 aircraft from industry, only 650 had been despatched to squadrons. Managerial and organisational changes were introduced that also had an immediate effect. Between January and March 1940 2,729 aircraft were produced by British industry, of which 638 were front-line fighters. However, between April to May 1940 aircraft production increased to 4,578 aircraft, some 1,875 of which were fighters. By June 1940, British fighter production reached 250% of German fighter production, whilst the overhauled repair service returned nearly 1,900 aircraft to action, many times more than their German counterparts. Consequently, German fighters available for operations during the Battle of Britain fell from 725 to 275, whilst fighters available for RAF operations increased from 644 on 1 July 1940 to 732 on 1 October.

Key to the success of the plan was the Directorate of Aeronautical Production which began work in March 1936 and had two goals: rapid expansion of defence industrial production and the dispersal of the defence industrial base to protect against air attack. By October 1937, there were five shadow factories already in production, whilst in July 1938 one shadow factory completed its first complete bomber. In time, the plan was also extended to industry in Australia, Canada, New Zealand, and South Africa.

The most famous of the shadow factories was at Castle Bromwich near Birmingham, which today is the home of Jaguar Cars. The plant opened in June 1940 and after some initial problems went on to build 12,000 Spitfires of twenty-two variants! The Shadow Plan also standardised development and production. For example, the Rolls Royce Merlin engine became the powerplant for many wartime aircraft. The plan also looked to the future by helping to fund the development of the jet engine and the world's second operational jet fighter, the Gloster Meteor, which entered service with the RAF three months after the German Me 262.

By 1944, there were 175 dispersed shadow factories in operation, many of which were linked to industries not traditionally associated with defence but with relevant supply chain expertise. The most famous aircraft to come from the plan apart from the Spitfire and Lancaster was the "wooden wonder," the de Havilland Mosquito, a twin-engine fighter bomber that could outstrip most single-engine fighters. The RAF was not the only service to benefit. The new King George V class battleships were built from 1936 on by many workers and technicians recruited under the Shadow Plan, whilst the British Army got new tanks some of which, contrary to popular myth, were not at all bad as well as the best field gun of the war in the 25 Pounder.

Much of Britain's defence, technological and industrial base has eroded since the end of the Cold War. Production facilities are few and orders even fewer and only seem to come when there is a political rather than a strategic imperative. Major systems only survive from cradle to grave because industry has learnt the vital need to tie government into contracts with punitive consequences when broken, whilst much of the "kit" ordered has more to do with industrial policy than defence policy.

Consequently, the unit cost of equipment British forces desperately need has become inflated, and much of it is obsolete before it is even fielded because innovation and technological advancement have been "de-prioritised." This has led to procurement disasters, including the notorious Ajax armoured infantry fighting vehicle, a platform that has had so many systems mounted as the requirement has been changed it looks more like a Christmas tree than an armoured vehicle. The programme has taken over thirty years to develop since the requirement was identified, at a cost of £3.2 billion, and not one vehicle is likely to be delivered until 2028 at the earliest. This sorry saga has led to some perverse adjustments. For example, the FV430 Bulldog armoured fighting vehicle will likely remain in service until 2030, sixty-seven years after it first entered service with the British Army.

The Ukraine War has also demonstrated the folly of emaciating Britain's defence industrial base. The Defence and Security Industrial Strategy is not yet a Shadow Plan and if it is not to be yet another of those "wizard wheezes" announced with much fanfare only to be lost in the vacuum of political irresolution it will need to be pushed through. It will also need to forge new partnerships across the entirety of a radically reconceived British and wider European security and defence supply chain that includes NATO's Defence Production Action Plan, the EU, governments, prime contractors, defence sub-contractors, systems-developers and providers who have thus far had little or nothing to do with defence. For that to happen the EU will also need to change its punitive Third Country rules which effectively excludes Britain from major contracts.

The Shadow Plan is the great unsung hero of the British war effort between 1935 and 1945. Without the plan Britain would have been defeated in 1940. Britain may not now be *at* war, but it is certainly engaged *in* war and, like 1935, it could soon find itself engaged in a systemic struggle if it does not move decisively to deter it. Such struggles are not won by fine words and crafted strategic documents that meet the political need of the moment. They are won by the sustained, systematic, and considered application of resources, technologies, equipment and forces over time and space.

PART TWO

THE RETURN TO STRATEGY

The enemy of a good plan is the dream of a perfect plan.

Carl von Clausewitz

5

A RETURN TO STRATEGY

We did not hit the crisis hard enough or fast enough due to a sclerotic London because we failed to apply the lessons from past crises and still do not.

Former Health Minister Lord Bethell on the COVID crisis,
7 December 2023

Domain Defence

The industrial age is giving way to the information age. Or to be more precise the digital-information age. Sound strategy in such an age is vital if London is to future-proof Britain so that risk can be managed and the British can help shape Europe and the wider world of the twenty-first century, and not be a hapless victim of it. How? First, as Winston Churchill once said, "The farther back you look, the farther forward you are likely to see." Second, Britain must once again have the political courage to consider the strategic worst-case. For example, what if the Americans at some pointed pulled out of NATO? Britain helped create structures, partnerships, coalitions, but above all institutions to deliver security and defence that were both geopolitically credible and affordable.

Critically, Britain's future defence will be domain-led so what is a military domain? The US Department of Defence defines the air domain as "the atmosphere, beginning at the Earth's surface, extending to the altitude where its effects upon operations become negligible."[1] The UK Ministry of Defence describes the importance of "the advancement and protection of the UK's national interests, at home and abroad, through the active management of risks and opportunities in and from the maritime domain, in order to strengthen and extend the UK's prosperity, security and resilience and to help shape a stable world." And in June 2016, NATO Secretary-General Jens Stoltenberg said that

> We [NATO Heads of State and government] agreed that we will recognise cyberspace as an operational domain. Just like air, sea, and land. Cyber defence is part of collective defence. Most crises and conflicts today have a cyber dimension. So, treating cyber as an operational domain would enable us to better protect our missions and operations. All our efforts to strengthen defence and deterrence depend on the right capabilities and the right resources.[2]

All domains are information-dependent, and every crisis has a critical information dimension, even if each domain has certain unique characteristics. The information domain should thus be seen as the medium through which strategy- and mission-critical information flows, defined by the bounds of utility where and when its effect upon operations becomes negligible. Like the air domain, the information domain is a contested space upon which there are clearly defined limits. The information domain also requires the application of control, the exertion of command (like the sea domain), the fostering of close co-operation, and the systematic and sustained exerting of pressure (like the cyber domain).

In other words, future war will be a weaponised information space. AI, machine-learning, quantum-computing, hyper-fast command and control will all be information-led, much of it automated. As such, information deterrence will be critical to structural deterrence, not least because of the hybrid marriage between human command and automated action, in which future systems and platforms will "educate" and "inform" each other almost instantaneously in pursuit of mission success.

The military information domain will be necessarily strategic and stretch far beyond any narrowly conceived battlespace, with virtually instant interoperability between Allied and Partner forces the key component of any defence. Defence policy, security policy and military strategy will thus be bound together in a seamless domain that enemies will repeatedly seek to penetrate and disrupt. Deterrence will thus demand the capability, capacity, and communication necessary to counter, interdict and retaliate against such incursions, with Britain at its core.

The hyper-fast nature of future war, as well as automated conflict in the grey zone short of war, will also be information-critical. Future hyper-fast war will thus need reconceived command and control as robotics, super-fast sensors, analytical logarithms, automated orientation, observation, decision, action, and assessment become central to the conduct of war. For democracies that imperative will create a difficult ethical dilemma that will likely prove less challenging for autocracies. Where exactly will the human commander sit in the command chain if any delay in the execution of war is a profound if not critical weakness? The answer will be both structural and architectural, with the information domain the glue that binds policy and strategy, deep intelligence, real-time intelligence, structural intelligence and awareness, human awareness, and artificial "awareness."

Power projection and people protection will thus be the twin pillars of future deterrence, both of which will be information-critical. Therefore, future deterrence and defence will depend on a robust capacity to project information and kinetic power together whilst also denying an enemy the capacity to disrupt the two great "orders" upon which credible deterrence and defence will stand—the order of battle and societal order. Ultimately, deterrence *is* information, with messaging critical to its ethos, method, and credibility—be it deterrence by denial or deterrence by punishment. A robust information domain will thus be essential for credible deterrence precisely because the capacity to operate to effect at speed on land, in the air, at sea, in space and in cyber-space will all depend on structural and real-time knowledge. Assured information will also be the glue which binds the citizen to the state and thus fosters resilience.

Enduring Principles of War

In 1925, Colonel J.F.C. Fuller elaborated his nine principles of war which can be thus summarised: clear direction, the need for offensive action, surprise, concentration of force, distribution of force, security, mobility, endurance, and determination. It is no coincidence that Fuller's principles were established in the wake of World War One during which radio revolutionised the speed and command of war. It is also no coincidence that Fuller was writing in Britain at the same time John Logie Baird was experimenting with television, and Robert Watson-Watt was experimenting with what eventually became radar. Both technologies concerned the development of media for the transmission of complex information instantaneously over ever-increasing distances, to be understood and acted upon by ever more sophisticated command-and-control systems, reinforced by an ever more immersive civil defence.

Fuller's nine principles might be a century old, but they were crafted for the efficient and effective use of force and resource in what was a new information age during which the Western way of war was being established. "Steel before flesh" has now become, over-optimistically and, perhaps prematurely, "technology before flesh."[3] Today, the world is entering a new digital-rich information age in which new technologies will be used for war and peace. These new technologies, both platforms and systems, are as information-hungry as policy and strategy. Artificial intelligence, quantum computing, machine-learning, robotics, and nanotechnologies will be at the forefront of peace and war throughout the twenty-first century and Britain needs to harness them. Indeed, they will be the quintessence of conflict in many instances, even if they will not be the end of conflict itself. The need for a balance between technology and flesh will be again vital for a power such as Britain, emphasising the enabling relationship between machines and brains, because for many years to come "tech" will enable human action, not replace it.

Fuller's nine principles have stood the test of time precisely because they assumed the existence of an information domain, albeit one that is far more primitive than today. Generating, securing, employing, and responding to information is already the *sine qua non* of command and control in warfare. Across the conflict spectrum much of that cycle of information is becoming necessarily automated, with sensors and algorithms informing artificial intelligence, precisely because it is information that is the change dynamic that could rapidly transform peace into coercion and manoeuvre war into hyperwar. John Boyd's OODA loop (observe-orient-decide-act), which explained how agility could overcome power, is being transformed into an ODAR loop (observe-confirm-autonomously, decide-swarm-autonomously, adjust and repeat—at ultra-high speed). Therefore, deterrence at the high end of conflict will increasingly depend on artificially, autonomously generated information simply because any other form of response would be slower and thus potentially fatal. In other words, it will require an all-of-state response.

Britain is well-placed to help lead Europe as it adapts to the changing character of war. Future war will also see coercion short of war through the exploitation of vulnerabilities, and the threat of systemic war, whilst the conduct of war itself will demand all national instruments of power—diplomatic, informational, military, and economic—both virtual and real. Harnessing such instruments to effect requires strategy so that they can be effectively used through structure. Grand strategy, the application of immense means in pursuit of high ends, ideally applies big power through integrated whole of government structures and mechanisms. Much of Britain's future ability to conduct effective grand strategy will concern the generation, speed of relevance, command, and application of information. In that light, Britain's contemporary information operations must be seen as just the beginning of an information revolution. For future grand strategy information should not so much be one domain amongst several, but a domain eminent. *The Oxford English Dictionary* defines "domain eminent" as "lordship of the sovereign power over property in a State, with right of expropriation."[4] The grand-strategic information domain will thus require lordship of the sovereign power of the state over strategic-critical information, with the capacity to exclude and deny adversary states and support allies and partners.

Future war and future peace will thus be information-dependent, but of course that does not mean that states will act intelligently with the information they have. Putin's decision to launch the Ukraine War was irrational in the extreme and once again revealed Abraham Lincoln's truism of history: power corrupts, and absolute power corrupts absolutely. It is also a war that was rendered more likely by constant denial over a significant period of the available information by so many leaders on both sides of the Atlantic—information which indicated that Putin would do exactly what he was threatening to do. President

Xi should also be taken at his word. He has repeatedly said it is Beijing's determination to reunify the Republic of China with the People's Republic by 2049 at the latest. There can be little doubt he will try with all the means available to him.

Where Can Britain Best Add Value?

In that light, and given the pressures on the Americans, where can Britain add value? It is reasonable to assume that even if the Americans one day abandon NATO, which remains highly unlikely but more possible than it once was, it would not also abandon its relationships with states it still regarded as like-minded. One area that would play to Britain's soft- and hard-power strengths is information deterrence. Deterrence *is* information: the strategic communication of grand strategy to adversaries and allies alike. Information deterrence would reinforce all other elements of deterrence and do exactly what it implies: deter "intelligent" warfare in which the primary enabler *ultima ratio regum* will be applied information. Information advantage should thus become one object of British strategy in an increasingly contested space and thus provide a central pillar of the future defence of not only Britain but all democracies, be they in Asia, North America, Europe or elsewhere. Equally, the virtual must be reinforced by the actual—covenants without the sword are but words and all that. Without the crucial underpinning of credible military force information deterrence would be little more than political spin, and Britain has already suffered enough from that.

"Future war" will not simply be war in the future, but also a new way of understanding ends, ways and means. As such, future war will range across the seven strategic and operational domains at the core of this book—air, sea, land, cyber, space, information, and knowledge. Information deterrence will demand not simply the mastery of all other domains, but the capacity to rapidly generate and act upon that most important of strategic commodities—understanding. In particular, the hyper-fast understanding of what is happening where to whom by whom and with what, and thus how best to respond. Britain is again well-placed to build such architecture.

The objective would be to generate the expected unexpected in which adversaries would regard Britain as an intelligent as well as a capable but essentially dangerous power which they expect to do routinely what they least want. British strategy should be devoted to the creation of a reinforcing defensive network-centric strategic architecture with the capacity to project decisive influence and deter one's adversaries across the information and cyber domains well before a shooting war begins. The purpose of such a defence would be the efficient generation and organisation of immense means via better understanding of their relevant utility in war. The shaping and conduct of warfare would thus be driven by the quality, extent and speed of the information afforded decision-makers and their capacity to respond. To operationalise such a paradigm (for that is what it would be) London would first have to establish two critical definitions central to the deterrence and conduct of war: domain and strategy.

A grand-strategic information domain would further necessitate significant changes to London's security and defence structures and resources if it was to credibly secure British critical interests and those of its allies and credibly deter adversaries. Future war in the twenty-first century will demand that Britain can deter not only through demonstrably intelligent applications of power, but also by being a demonstrably intelligent warfighter across the civil-military spectrum. To such an end, Britain would have to design and develop a new security and defence community from both within and outside government, from the warfighter to the thinker, which could rapidly disseminate and apply needed actions from available information.

To design such a strategy would also require once again a profound understanding of adversaries at all levels of command, structure, and society and how they conceive and plan to conduct future war. This was a capacity that London once took for granted and should be able to do so again given the potential of British academia and expertise. Equally, the generation of the necessary information to be judged would also depend upon the extent to which AI, machine-learning and big data can also be harnessed to enable rapid analysis and decision-making.

Information and Disinformation

Given that information will be the indispensable commodity in future war it will also be locked in a constant struggle with disinformation. Securing assured information will thus be a fundamental principle of war. For example, strategic *maskirovka* (deception) is one of a trinity of "mass" elements Moscow is developing in its future war strategy—mass disinformation and deception, mass disruption and mass destruction. Strategic *maskirovka* is specifically designed to exploit the many seams that now exist in Western societies by manipulating public opinion to blunt any meaningful policy response to Russian coercion. Moscow's strategy is being implicitly aided and abetted by some European governments, including Britain, the vulnerability of which to Russian interference has been revealed by the Russo-Ukraine War. When the Americans put their National Planning Scenarios online some years ago several European states demanded they be taken down for fear of revealing just how vulnerable they had permitted their states to become.

David Richards is clear:

> I have said for many years that we should not be surprised at Russia's actions. Russia is simply doing what Russia has always done when opportunity arises, and deterrence fails. We've got to learn to do the

same, but BETTER. In this case, using the Russians as a vehicle to a better understanding of what is needed.

The evolving Russian view of information warfare is also intimately linked to their concept of future war, whatever the failings of the Russian campaign in Ukraine. Moscow's use of the information domain is reflective of the increasingly coercive and subversive role of information across the conflict spectrum. This further reflects a broad concept of future war that stretches from cold peace to hot war. For the Russians, use of information and disinformation is nothing new. *Maskirovka* was used extensively during World War Two to confuse German commanders, with perhaps its most notable application during the Battle for the Kursk Salient in 1943.

For a state such as Russia, which already sees itself engaged in "perma-war" with Britain and the West, disinformation is central to an information shock gambit. It is part of permanent low-level war, with a clear escalatory capacity to high-end conflict by fostering chaos in adversary states. Why is Britain a particularly important adversary for the Russians? To the Russian mind Britain is a metaphor for the United States and can be attacked without the risks associated with attacking the Americans. Moreover, France and Germany are still seen by Moscow as potentially malleable, whilst the rest of Europe simply does not matter in the Russian strategic calculation. Britain thus has an outsized presence in Russian grand strategic thinking, the more so since Britain took the lead in support of Ukraine in February 2022.

That is why London needs to face strategic reality. If war usually involves trading space for time, future war will create chaos to offset weakness. Russian statecraft has no ideological goals other than the preservation and furtherance of power for a corrupt elite via three conduits: complete control of governance; preservation of their own power and wealth; and undermining foreign enemies. Strategic *maskirovka* and its partner *dezinfor-*

matsiya (disinformation) are simply grand-strategic adaptations of Russia's traditional use of battlefield deception. *Maskirovka* is thus war that is short of war, part of a purposeful strategy that combines the threat of massive force with disinformation, destabilisation, disruption, and deception. It is a permanent coercive strategy conducted from the very top of the Russian state, through multiple and repeated messaging and often through deniable media. By implying escalation to mass destructive force, the aim is to keep adversaries politically, socially, and militarily permanently off-balance.

Information deterrence will also be central to the increasingly intimate and vital relationship between power projection and people protection. In certain respects, autocracies are less vulnerable to information exploitation than democracies because it is easier to undermine the social and political consensus upon which deterrence and defence must necessarily rest in the latter. For Britain a persistent challenge is how to make the perceived weakness of its open society a strength. Information redundancy and resilience will be key: social media, open-source intelligence and commercial agility will all need to be harnessed to act as information multipliers. Hitherto, London has existed in a naïve world in which openness is an end.

Britain does have comparative advantages over Russia. London can rapidly upscale its security and defence effort if it builds vital strategic public-private partnerships to make Britain's critical information and energy infrastructures more resilient. The December 2023 UK government Resilience Framework did not do that. Rather it did what "strategies" so often do when London has been caught off-balance: fail to come up with a quantifiable plan to invest in something vital, and which London itself sees as vital. Rather, the concept of "partnership" is too often seen by London as a mechanism to shift the cost of enhanced resilience from the public purse onto the private sector.[5] A real information partnership between the state, the private sector and its citizens would be state-led and would require a reconstructed civil defence mechanism in which the private sector generates redundancy and capability. Sometimes the lack of strategic thinking in London borders on the absurd. Calling upon people to stock up on candles and hand-driven radios should the internet crash in an emergency (it will) is almost as bad as the advice given by London to families in the 1960s to "Protect and Survive" by hiding under the kitchen table in the event of a nuclear attack. Hit the enemy where it hurts! Autocracies by their very nature are more vulnerable to information decapitation and need to be made to understand the dangerous consequences of their own aggression.

Hard Information Power

Fuller's nine principles beg a further question: why should information be seen as a military domain in much the same way as air, sea, land, cyber and space? It is the seamlessness of action future war will demand that makes the need for a bespoke information domain. Fuller built on Carl von Clausewitz and his notion of economy of force: the application of combat power to the main effort in the most effective and efficient manner to limit the use of such power for subsidiary aims. The very judicious use of force implicit in Clausewitz's theory was necessarily information-led. A military grand strategy for a country such as Britain today would also necessarily be built on economy and utility of force, which will be vital for credible deterrence given the scale of potential conflict and the sheer size of the theatres and domains in which it would be fought. The very idea of "war" will also include manipulation and coercion using both digital and analogue civilian-led means allied to a grand-manoeuvrist distribution of forces, all of which would be dependent on distributed information, both real-time and structural.

Information advantage will thus become the most sought after commodity in both peace and war, but it will need to be consciously designed, developed, and deployed, just like forces in any other domain. Like any other domain, such will be its value that it must also be protected, assured, and denied. In other words, *si vis pacem, para bellum*!

Five Eyes or More Eyes?

The Five Eyes intelligence group of America, Australia, Britain, Canada, and New Zealand is a vital information-age hub upon which British strategy has relied for eighty years. Five Eyes also reinforces Britain's essentially Euro-Atlantic focus with a global security and military intelligence perspective.

However, to properly exploit Five Eyes in its ninth decade a politically and bureaucratically sclerotic London will need to become far more agile and offer far more. Five Eyes shares intelligence, analysis, technology, tradecraft, and statecraft like no other grouping, built on a shared history, worldview and strategic culture that is often lacking in bilateral or institutional relationships—a culture of trust. The Five Power Defence Arrangements have been built up over eighty years of sharing the most sensitive secrets since the UK-USA Agreement began in 1943, with Britain very much in the lead through the British-US Intelligence Communications Agreement. Five Eyes began as a collaboration to penetrate Soviet capabilities and intentions. After two decades of a counterterrorism focus the now very US-dominated Five Eyes is adapting once again to face the peer competitor challenge of China and Russia. It is also adapting to meet the challenge that quantum computing will provide for tradecraft, the pooling and sharing of raw intelligence and data generation, workshopping analysis, thematic and threat reporting, as well as the operational conduct of operations, as the sheer scale of data increases exponentially across the HUMINT (human intelligence) and SIGINT (signals intelligence) spectrum.

The annual gatherings of Five Eyes leaders will take on ever more importance, as will the already extremely tight relationships between the respective national security advisors of the five countries. Britain also faces a dilemma. Much of the success of Five Eyes is due to personal relationships between key personnel and the very real trust in each other that exists. This has increased the level of mutual dependence, even if at times non-US partners find the American NOFORN ("Not for release to foreign nationals") onerous—especially when they have generated the raw intel. However, unprecedented access to unrivalled American assets and capabilities has a value far greater than the cost of having to deal with the US inter-agency morass.

If Five Eyes should become more eyes who else could be invited? Japan is already a virtual sixth member but the very nature of how Five Eyes functions makes it hard to expand further. At the same time, given the centrality of information-led deterrence and defence, particularly to NATO, Five Eyes has a vital role to play in the wider defence of the West. Five Eyes has also long generated jealousy from France, Germany, South Korea, and others, all of which have expressed an interest in joining. The challenge is that to function the network must not become overly vulnerable to so-called "inherited threats." Five Eyes has suffered from a breakdown at times of "insider trust," most notably because of Edward Snowden. Those in the network are also increasingly concerned by the threat posed by cyber penetration, which is very hard to stop. The future may well be hybrid Five Eyes-Plus arrangements that place particular responsibility on London, and by which trusted third parties enjoy elevated access to the network, as they do to a significant extent already.

For Britain, a Five Eyes future will also mean accepting significant change to the UK-US Special Relationship. The key to the cohesion of a necessarily broader coalition of like-minded nations the world over will be a new kind of alliance-partnership built precisely on assured information sharing. For such a grouping, access to US (and British) intelligence products will no longer be seen as American largesse (or to a lesser extent British largesse), but rather in the interest of both to share intelligence products whilst still protecting both methods and sources. Therefore, Five Eyes should be expanded into a Five-Eyes-plus-allies-and-partners network, a new network created for the security and defence needs of today. The importance of democratic allies and trusted partners across the world (the alliance-partnership) having as much access as possible to a shared strategic and operational picture is undeniable. Such access will be critical to decision-making in a fast-moving, contested environment across a complex conflict spectrum, which ranges from terrorism to high-end warfare and in which time and distance matter. There will be many difficulties expanding Five Eyes because of its nature, the complexity of US inter-agency relationships, the nature of intelligence and those who work within it, institutional inertia, the potential loss of "special" access to US products for the four other members of Five Eyes, the risks of revealing sources with a wider network, and the abiding issue of mutual trust. However, it needs to be done and Britain must push for it.

Five Eyes is rooted in an analogue world. Over time Five Eyes changed from an exchange between trusted peers to privileged access for smaller powers to US signals and human intelligence assets. Now, in a digital world in which the threat straddles the real and virtual worlds, and as US forces and resources become ever more stretched and the scope of threat becomes global, a broader array of more capable allies and partners to see the same intelligence products is in Washington's interest. For that to happen Washington would need to stop seeing access to intelligence as a way of disciplining allies and partners but focus rather on the better and greater sharing of products whilst still protecting sources. Allies and partners would need to take the provenance of such products on good faith—not something the intelligence community finds at all easy.

The current multi-level system of access generates patent absurdities which could be a potentially critical vulnerability during a crisis. The NATO Secretary-General cannot access either US NOFORN or Five Eyes intelligence. Non-American NATO commanders of US troops in the field are excluded from mission-critical intelligence open to their subordinates. The first highly respected German NATO Assistant Secretary-General for Intelligence and Security was denied access to US and Five Eyes products. Such friction can lead to a delay in Alliance action of days not hours.

To generate and maintain cohesion across the new alliance-partnership and to penetrate the decision-cycles of adversaries in a very different world to the one which spawned Five Eyes, a new privileged group is needed based on a hybrid system of access somewhere between NOFORN and Five Eyes' openness. This new system would include all the NATO allies on a probationary basis, plus the NATO Asia-Pacific partners (AP4), Australia, Japan, New Zealand and the Republic of Korea. Such change would impact on the so-called UK-US Special Relationship, much of which is intelligence-based, and there would be elevated risks of insider trust, source protection and cyber penetration, but such risk should be manageable given the improved unity of purpose and effort it would generate.

The Russo-Ukraine War revealed the dangers of different allies and partners having access to different grades of intelligence. For example, the US was working to a different level of intelligence than its four Five Eyes partners, which helped generate very different interpretations and analyses. Ultimately, enhanced deterrence, defence and security can only be achieved in the twenty-first century via a tight coalition of strategic effort, dependent in turn on the right intel getting to the right people at the right time.

6

THE UTILITY OF (BRITISH) FORCE

Armed conflict is a human condition, and I do not doubt we will continue to reinvent it from generation to generation.

General Sir Rupert Smith, *The Utility of Force*

A Little Bit of Everything...

What are the British armed forces for? In 2023, London assigned four missions to Defence as part of a "Strategic Framework" which captures the mismatch between ambition, capability and capacity from which Britain's armed forces suffers. These missions are as follows:

> *Shape the international environment.* Defence contributes through: Global approach to campaigning and competition; alliances and partnerships; bilateral relationships and multilateral and minilateral [whatever that means] groupings; engagement with middle ground powers; supporting others to deliver their security; an integrated global network of people and bases.
>
> *Deter, Defend and Compete across all domains.* Defence contributes through: Credible capabilities, nuclear and conventional, cyber and

space; our role in NATO; our support to Ukraine; a resilient underpinning of stockpiles, enablers and intelligence.

Address vulnerabilities through resilience. Defence contributes through: Defence of the homeland; protection of airspace and critical national infrastructure, including sub-surface; support to the civil authorities; economic security; use of our Reserves.

Generate strategic advantage. Defence contributes through: Our people; a strong relationship between Defence and industry; slicker acquisition processes; modernisation through innovation; digital and data, science and technology; our role in supporting economic growth and national prosperity, including defence exports.[1]

Given the state and size of the British armed forces the Defence Command Paper is the wish-list of wish-lists. It also reveals the culture of political spin central to British defence "strategy" and why the British armed forces have a little bit of everything, but not much of anything. So thinly is a very small force spread across so many missions that the armed forces are reaching a tipping point at which the £50 billion or so London invests in defence is simply no longer value for money. Therefore, if London wants more value for money through more efficiently matching ends, ways and means it is going to have to spend more not less on defence. In February 2024, it was announced that all capital spending on defence had been suspended, placing all longer-term projects in doubt, though these are vital to London's own Pollyanna defence strategy.

Such inconsistencies were an important reason why Sunak announced the hike in defence expenditure to 2.5% GDP by 2030 with year-on-year increases in funding to meet that target. However, this episode also highlights again the inherent tension between short-term politics and longer-term strategy from which the British defence effort has suffered for far too many years. It also underlines the extent to which London has retreated from strategy and the vital need to apply consistent power consistently over time to secure Britain's critical national interests.

...but not Much of Anything

As ever, it is the armed forces themselves who are endeavouring to close what is an impossible gulf. In September 2020, the British established the Integrated Operating Concept (IOpC) with the aim of controlling the conditions and tempo of military-strategic events, rather than being forced to respond to adversaries or even allies. The IOpC is the core of contemporary British national strategy. Now that "efficiencies" have pretty much been exhausted "integration" is the new buzzword. The IOpC thus seeks to integrate all the national instruments of power: diplomatic, informational, economic, finance and trade policy statecraft, along with bedrock "credible" military power. Therein lies the problem. The ability to deter war remains central to British military purpose, but such deterrence must now also be credible below the threshold of war itself. That is why Defence Command Paper 2023 calls for a decisive shift in the British military approach to warfare with the establishment of a new strategic approach that both mobilises the existing force to meet today's challenges and drives its modernisation, which is vital if it is to retain any utility beyond the most permissive scenarios.

Given the IOpC, the pressing challenge for the British is to sustain the vital relationship between the defence operating model, the defence planning assumptions and force structure. The problem of planning assumptions has been a thorny one since Strategic Defence Review 1998 when an effort was made to understand the likely minimum scale of effort Britain would need for both home defence and expeditionary operations that were significant in size, reach, fighting power and staying power. The assumptions were soon broken by Afghanistan and then finished off by Iraq. Defence planning assumptions establish a level of force requirement over and above day-to-day military tasks and are usually divided into small-, medium-, large-, and full-scale. The scales of effort are supplemented by readiness, endurance, and concurrency levels, and the higher up the scale or levels a force the more expensive it is to maintain.

The level of readiness is critical because this is the notice period within which units must be available to deploy for a given operation, whilst the time an operation is likely to last is known as endurance. Concurrency is the number of operations a British force can undertake at the same time, given the scale of effort and duration. The tensions at the heart of the defence planning assumptions led inexorably to joint-force concepts being developed, first with the so-called Joint Rapid Reaction Force (JRRF), and today with the Joint Expeditionary Force (JEF). David Richards writes,

> The JEF was as much a political military construct as anything else and meant to provide a forcing function within NATO through closer cooperation between the more innovative and dependable northern members of the alliance. Whilst it has grown, it is still not capable of fighting as an independent Joint Force. It needs NATO enablers and NATO command and control. In practice it is a grouping of like-minded nations who have agreed that they would be happy to operate together through and as part of NATO.

Britain's military-strategic dilemma is that however deeply integrated and joint its force, it can now only realise the most permissive of its defence planning assumptions without the support of allies and partners. In other words, joint forces must be bolstered by combined forces at the price of political autonomy. However, projected collective security missions, as opposed to existential collective defence, are intensely political and lack the automaticity and consistency of action defence requires. Successive defence reviews since 2010 have tried to fudge this hard reality. Strategic Defence and Security Review 2010 (SDSR 2010) published endurance and concurrency details for its vague Future Force 2020 concept but little else, whilst SDSR 2015 conveniently left many defence planning assumptions unexplained and was very much the product of the "how much threat or force can we afford" school of defence-strategic thought. There was also a profound disconnect evident in its claim that a British expeditionary force would be, at its largest, 50,000-strong, which was some 30,000 bigger than envisaged in Future Force 2020. There was also no proper analysis of the type, quantity, and duration of operations Britain's armed forces would need to be sized and shaped to conduct. In other words, both SDSR 2010 and SDSR 2015 were more accounting exercises of the moment than strategies designed to stand the test of time and were thus subject to an invasion by a changing reality.

All of London's recent strategies suffer from the same tensions. There are few if any defence planning assumptions for the use and deployment of London's new rhetorical flagship, Integrated Force 2030, but a host of very demanding and resource-rich tasks and commitments that range from high-end warfare, long-term engagements abroad, crisis response, defending Britain and its overseas territories, as well as the maintenance of a continuous at sea nuclear deterrent. Therefore, it is hardly surprising that since 1998 not one of the five-yearly future force visions has been delivered on, which results in the strange position in which London finds itself today: claiming to be ever more militarily agile and capable and yet ever more reliant on the Americans and NATO, while promising to do more for both. Britain's defence strategy is wholly dependent on NATO realising its New Force Model under the current NATO Defence Planning Process—and that, to say the least, is highly uncertain.

The NATO Option

After a thirty-five-year post-Cold War hiatus NATO is again the lodestar for British defence strategy, and it is vital that Britain implements in full its commitments under SACEUR's (Supreme Allied Commander Europe's) "Family of Plans."[2] The Concept for the Deterrence and Defence of the Euro-Atlantic Area (DDA) is at the core of the effort, with the spearhead the new Allied Reaction Force (ARF). It is no exaggeration to suggest that given the political mood in the United States and the global overstretch from which US forces suffer, realising SACEUR's plans might be the last chance saloon for the European allies. Britain needs to be front and centre of that effort. Unfortunately, the fact of NATO too often produces a "we can no longer really afford to do this or that, so our allies will have to do it, but don't tell anyone else" mindset in London. The problem is that every other European ally operates from the same viewpoint, with NATO now an "after you please, America" alliance. NATO is thus increasingly like one of those Soviet propaganda movies of old which were all façade and no substance and in which "cohesion" is presented as a meaningless alternative to capability and capacity.

For Integrated Force 2030 to be credible it must at the very least mirror US efforts to modernise its forces by moving away from a focus on counter-insurgency operations and back to high-end power projection, and yet retain sufficient mass to undertake and sustain a range of missions. Again, the most appropriate organisational model for the British to follow is the US Marines Corp which is an intrinsically joint force. However, for the British force vision to be realised would require a level of unity of purpose and effort amongst Britain's defence chiefs hitherto unknown, and a willingness on their part to speak hard truth to political power, and with one voice. There would also be hard choices. If the nuclear deterrent is taken as a given and assuming cyber defence (and offence) would be as much a civilian as a military cost, the centre of gravity of any such vision must be a future, US-interoperable high-end British joint expeditionary force. Whilst the US Marines Corps is the most appropriate organisational model, the force structure would be more akin to the new Allied Reaction Force and in time an Allied Heavy Mobile Force. This would place Britain firmly at the centre of the NATO force and command structure.

The strategies also suggest a future joint British force will be supported by emerging and disruptive technologies. Again, the US Marine Corps is the most obvious parallel to the British armed forces, not just because of its comparable size (the US Marine Corps has some 182,000 active personnel supported by 38,000 reserves; the UK armed forces have some 149,000 active personnel supported by 44,900 reserves) but also because the US Marine Corps and the British armed forces are power-projection forces, with both increasingly focused on admittedly vulnerable carrier-enabled power projection. Such is the partnership with the US Marine Corps that the Americans operate F-35s from Britain's two aircraft carriers, HMS *Queen Elizabeth*, and HMS *Prince of Wales*, even if that has much to do with London being unwilling to buy enough F-35B aircraft to equip the ships properly.

Both the British armed forces and the US Marines Corps share other "virtue-out-of-necessity" attributes relevant to Integrated Force 2030. London's abandonment of a continental strategy and the centrality of the nuclear deterrent in its defence strategy has led inexorably to a kind of rough military logic about the nature of future interventions. First, any such posture precludes the kind of mass force that would be needed to fight a high-end war with the likes of China and Russia. Second, the focus is on a small but high-quality, deep, joint "strategic raider" force led by a single strategic command. Third, given the relatively small size of such a force and its "lightness," it assumes a high degree of interoperability with the US and Allied armies and access to US enablers to ensure follow-on mass and support.

Britain and the Allied Heavy Mobile Force

What does Britain offer NATO today? To deploy a force of 20,000 to support NATO Exercise Steadfast Defender in early 2024 London had to strip out almost all available troops and combat support from the existing defence establishment. What Steadfast Defender demonstrated is that if an Article 5 contingency took place that demanded two fielded and sustained divisions Britain would fail. Therefore, if Britain really wishes to remain a tier-one military power the very least London should be able to offer the Alliance, should the Americans be busy elsewhere, would be a new British Army built on at least eight warfighting brigades. This would be placed at the core of a twenty-first century "heavy" enhanced NATO Response Force (eNRF) that combined high-end first responders with follow-on forces. In other words, something akin to an Allied Heavy Mobile Force (AHMF).

It is also Britain's interests that in time the Allied Reaction Force evolves into an Allied Mobile Heavy Force with the British at its core. Therefore, to ensure NATO continues to be supported by the Americans Britain should move to ensure at least 50% of all Allied capabilities at all levels of conflict are European by 2030. To realise such a commitment, London should propose to Berlin and Paris the creation of a first-responder Allied Future Force able to act from seabed to space and across the multi-domains of air, sea, land, cyber, space, information, and knowledge. The AHMF would need to be sufficiently robust and responsive, and held at a sufficient level of readiness, to meet all threats to the territory of the Euro-Atlantic Area in the first instance, whilst also having sufficient capacity to support those front-line nations facing transnational threats, such as terrorism.

The eNRF would build on the Very High Readiness Joint Task Force (VJTF) and the NATO Response Force (NRF), as well as those very high readiness forces that will emerge from the NATO Readiness Initiative. It would also enable NATO to better exploit emerging and disruptive technologies and maintain a high degree of interoperability with fast-evolving US forces. Finally, it would act as a vehicle for the introduction into the Allied Order of Battle of artificial intelligence, super computing and quantum computing, big data, machine-learning, drone swarming and autonomous capabilities (e.g. manned-unmanned teaming, decoys, relays, and networked autonomous systems), and hypersonic weapon systems to enable an Allied capability to engage in hyper-fast warfare. This capability will be crucial to deterrence in the future.

The AHMF would also act as a vehicle for greater European strategic responsibility built on relative military capability and capacity and must be seen as such. Together with enabling combat support and combat service support, the AHMF would also need to be deployable in several guises and under more than one flag, including as a NATO-enabled European coalition (both EU and partners), or as a framework for coalitions of the willing and able. As such, the AHMF would be the natural development of both the Combined Joint Expeditionary Force and the Joint Expeditionary Force both of which would be part of the new heavy force.

Again, and it is worth re-stating, to ensure the AHMF is stood up in a timely manner the European allies, together with Canada, must by 2030 invest sufficient resources to ensure they collectively meet at least 50% of NATO's minimum military requirements identified by the Strategic Commanders, including fully usable forces required for covering the whole spectrum of operations and missions, as well as the strategic enablers required to conduct multiple demanding large-scale and smaller-scale operations. Such operations will be conducted both alongside US forces in a variety of regions inside and outside SACEUR's Area of Responsibility, as well as autonomously when agreed.

The political benefits for London of such a capability would be clear. First, Britain would still be able to exert leadership within the Alliance by pulling its weight and matching words with deeds. Second, such a "command hub force" would also enable non-US allies to plug into UK-led coalitions if the US was busy elsewhere. Third, it would enable the French and the Germans to buy into a new British commitment to European defence, even if it did require elements of Integrated Force 2030 to be kept at significantly higher readiness for longer than London would prefer. That, after all, is the price of being a tier one military power in the 2020s. Anything else is strategic and defence pretence.

It is important to remember the aim of British strategy is to deter not to fight the likes of Russia, let alone China and Russia together. Preserving peace or preventing war is the stuff of British strategy and to achieve that the West, with Britain to the fore, needs to be palpably strong enough. London must also consider the worst-case, which is a war in which China and Russia succeed in splitting the Allies, forcing the Americans to concentrate overwhelmingly on the Indo-Pacific, whilst weak Europeans find themselves facing reenergised and galvanised Russian forces in what would be a coordinated attack by an updated and much more dangerous version of the Axis Powers of old.

Defence Strategic Assumptions Beyond 2030

Maintaining tier one status beyond 2030 will mean nothing short of a radically different approach to British defence strategy. By 2035, everywhere will be a battlefield and everything will be a weapon. The Russo-Ukraine War has shattered hopes for peace that have existed since the end of the Cold War and the fall of the Berlin Wall. The post-Cold War world has come to a crashing end. There is now an urgent need for clarity concerning so-called "red lines," not only over possible use of nuclear weapons, but cyber warfare, information warfare and the use of emerging and disruptive technologies across the defence, information, military and economic space.

Emerging and disruptive technologies entering the battlespace could well revolutionise warfare by 2035, and at the very least profoundly change its character. Whilst Russia is the immediate threat the danger Moscow poses is also a consequence of Russian decline and an inability to adapt to the twenty-first century. This is why it is so dangerous. Instability in many parts of the world must also be engaged. Going forward China poses the greatest systemic threat due to a combination of great and developing economic and military power, an autocratic leadership that divides the world into adversaries and client-states, and rapid technological advance that is fuelled by industrial, cyber, and military espionage. China's rapidly ageing population is reinforcing Beijing's technology drive across the information, military, and economic domains. China and Russia must be systematically prevented from accessing defence-sensitive programmes, including dual-use technologies. Far greater efforts need to be made collectively to block Chinese and Russian industrial and cyber espionage. This also requires mitigating and closing vulnerabilities to authoritarian states by reducing reliance on their energy sources and other raw materials and eliminating one-sided dependencies in trade with China.

Deterring future war starts with properly learning the lessons from the war in Ukraine, but needs far more imagination than those lessons alone. There is a gaping hole between rhetoric and reality across the conflict spectrum, poor integration of policy and effort across diplomacy, information, military and economic domains and little consensus or even idea about how to proceed either nationally or collectively. The role of resilience as deterrence by denial will gain more importance. This requires boosting societal and democratic resilience. Vulnerabilities in democratic, social, economic, and political systems tempt and even invite intervention by authoritarian regimes, as has been observed in the last decade.

Maintaining sufficient threat-led defence investment will prove challenging going forward because of pressures on many sectors of society and due to structural economic crises caused by the financial and banking crisis, the pandemic, and the war in Ukraine. If democracies are to deter future war, leaders will need to understand and appreciate the value of defence and not just its cost, and communicate that to their respective populations.

Preserving a just peace and effective deterrence is the core business of the democracies in general, and NATO in particular. Deterrence will continue to be centred on conventional and nuclear force but must be reinforced by a new concept of deterrence that stretches across the hybrid, cyber and hyperwar mosaic. Transatlantic deterrence will be most effective if produced and implemented collectively. "Better together" will remain the best principle for deterrence and defence. It should operate on the entire spectrum ranging from nuclear and conventional to unconventional deterrence. Not everyone needs to do or be on board with everything that is required, but a common, integrated approach to competing and contesting is vital.

Decision-making will also need to be far faster if deterrence is to be founded on the speed of relevance because of the increasing prevalence of artificial intelligence, machine-learning, quantum computing, loitering strategic glide systems, deep-strike hypersonic missiles, intelligent and unintelligent drone swarms, and nanotechnologies.

Therefore, by 2035 at the very latest, Europeans will need a high-end, first-responder force that can act from seabed to space and across the domains of air, sea, land, cyber, space, information, and knowledge. Britain will be expected to be both a cornerstone and capstone power of such a force, which will be capable of sufficient manoeuvre to be able to respond to, sustain and exploit the fight from the Arctic to the Mediterranean, especially should the US be engaged in strength elsewhere.

Future major crises and wars will have a strategic, if not global, dimension, and will be politically very complex and militarily very demanding. In particular, the military use of artificial intelligence—such as in new generations of sensors, space-based capabilities, autonomous weapon systems, much-improved air and missile defence, drones, and long-range precision missiles—by NATO and its adversaries necessitates the ability to take urgent decisions in a crisis and immediate decisions in a war at both the political and military strategic-operational level. Precision and speed will be of the essence.

NATO will remain the premier organizing institutional and operational structure for Western security. Other frameworks, however, such as the US Unified Command Plan and Combatant HQs, the French-British Combined Joint Expeditionary Force, the European Union and the UK-led Five-Power Defence Arrangements with Australia, Malaysia, New Zealand, and Singapore will also play important roles in coalition operations and Britain will need the flexibility to exploit such groupings.

Defence of the Euro-Atlantic area is the defining requirement for NATO forces. It requires a capacity to concentrate forces and fires to counter Russian advantages. Defence in the Indo-Pacific region mandates a capacity to disperse forces and fires over long distances. Combining the two requirements will impose greater flexibility on NATO forces, such that they can operate within and beyond Europe, even if not under NATO command. At the same time, the United States is encouraging greater cooperation and force integration among Australia, Japan, Singapore, South

Korea, and Thailand, as a substitute for bilateral defence relationships that do not match new regional security dynamics. Therefore, there is today a greater degree of convergence between Euro-Atlantic and Indo-Pacific defence requirements from different points of departure. However, the latter will likely require more radical levels of innovation in the design of future forces.

Ultimately, deterrence of future war will require a profound reimagining of statecraft, with deeper synergies forged between policy, strategy, civilian and military force and resource and a new balance between power projection and people protection. Such synergies will be critical for the future functioning of NATO, but must be extended to all the democracies in what is a global emergency. Now is the moment to begin constructing such architecture because preserving the future peace will demand nothing less.

Future war and deterrence are not simply a question of technology. Democracies today are facing complex strategic coercion via applied disinformation, deception, destabilisation, disruption, and the threat of actual destruction. Democracies thus face a potential *Dreadnought* moment and digital decapitation through a combination of hybrid and cyber allied to new technologies being applied catastrophically by enemies against vulnerable, open societies.

China is "tomorrow's fight" and Russia the immediate fight. Together, China and Russia have embarked on a systematic strategy to exploit American weaknesses and vulnerabilities both before a conflict and in the event of conflict. Such measures include interfering with the American political process and seeking to exploit US military over-stretch. China is in a "waiting game" with the democracies, with Beijing firm in belief that in time the "correlation of forces" will favour China where its vital interests are at stake—Taiwan and much of Southeast Asia. Russia's decline means Moscow has no time to waste.

There are now apparent limits to US engagement in NATO even if the Americans remain fully committed to the Alliance. US military overstretch is being exacerbated by European military weakness even though efforts are being made at mitigation through the more equitable sharing of burdens and risks within the Alliance. To that end, a culture of worst-case-scenario planning and exercising must again be established. In any major future war, the democracies could face multiple simultaneous contingences via a series of global "feints" across technical domains and in the Indo-Pacific, Middle East, Black Sea, Arctic, and Europe.

Where to Begin?

A joke in the Royal Navy goes like this: a group of journalists are visiting HMS *Queen Elizabeth* and one of them sees the Windows XP icon on a computer screen. At the end of the visit, the journalist asks the commander is it true they use Windows XP? "No," says the commander, "we have nothing on board that advanced." The *Vanguard*-class ballistic missile submarines really do use Windows XP as part of the fire control system.

The importance of savings and efficiencies must not be a reason for yet more cuts, but rather to make investment more effective and targeted. To do that London needs to focus on what one senior officer called the "connective tissue." Interestingly, the three intelligence agencies, GCHQ, MI5 and the Secret Intelligence Service, have already done that. To be effective such an effort must have a defence vision that in Britain's case must reach out to at least 2035 and beyond. Only then will it be possible to identify what to change now, over the medium-term and the longer-term. The strategy would also need to consider how best to generate comparative advantage through the enhanced role and utility of people and the linking "glue" between them, as well as the exploitation of new concepts and technologies, such as the "Combat Cloud." That, in turn, would require a culture of experimentation to be properly established, recognising that all such experimentation involves cost. Not every experiment works. Above all, such a new approach will require imagination which is self-evidently lacking today.

Trevor Taylor and Andrew Curtis capture the challenge succinctly:

> [S]uccessive policy reviews have allocated a broad range of missions and tasks to the country's armed forces while, at the same time, making only limited resources available. In this case, there is a need to generate a coherent pattern of prioritisation across defence ... [T] echnological advances, especially in electronics and computing, which have transformed surveillance and communications, are putting the ability to generate and exploit information at the centre of military capability. The systems that generate, move, analyse, and protect information are prevalent in all five operating environments: on and under water, on land, in the air, in space and in cyberspace.[3]

In other words, whilst the cost of maintaining credible defence and deterrence increases the task-load, defence investment simply does not keep up.

The causes are manifold: defence cost inflation at as much as 8% per annum; the inability to build equipment the force needs and the incapacity of British industry to produce it when it is needed due to small production runs; the cost of in-life upgrades; the cost and pace of research and development; the additional costs of *juste retour* job-sharing in multinational programmes; the committed cost of pensions and other overheads; as well as the use of so-called "big ticket" programmes for political rather than defence-strategic reasons, such as creating jobs in Scotland, evidenced by the number of "one-hit ships" in the fleet and Royal Fleet Auxiliary. There are also a host of costs built into defence due to outsourcing and the bureaucratic desire to keep as many defence costs as possible off the balance sheet. For example, the Royal Fleet Auxiliary is paid at civil service rates but uses a commercial structure that greatly inflates personnel costs. At the same time, even though the pay rates are commercial they are not high enough to attract enough people to crew Royal Fleet Auxiliary ships.

At the very least, London must take back ownership of the defence and technological imperative. A great deal of unnecessary cost has been caused by the pretence of competition in defence procurement which inflates project costs as soon as a contract is signed. This is made worse by incompetent project management on the part of the Ministry of Defence, even as there is also a vital need for the MoD to take back central control of critical processes, such as defence technology development. Too often unnecessary cost is also driven into equipment, as one senior officer explained. The "risk automated system" for nuclear-powered submarines has caused very significant additional cost because the definition of acceptable risk is now so low as to be beyond the extremely unlikely. This London-driven extreme safety culture is understandable—Britain's submarines are nuclear after all—but like so much in British life these days such abundance of caution is making it impossible to afford new hi-tech equipment.

Industry also sees the MoD as a "parasite on industry," with a culture far removed from the incentive and innovation-driven culture of business. This must be quickly fixed because as the development of software and systems accelerates, the relationship between systems and platforms will become ever more complex. For example, the life cycle of two aircraft carriers is planned as fifty years, but the life cycle of some of the systems on them will be measured in months. Consequently, the systems HMS *Queen Elizabeth* will carry at the end of her in-service life will be very different from the ones she carries today. The best that can be said for MoD procurement policy is that it is very good at producing yesterday's equipment tomorrow. The new frigate HMS *Glasgow*, which is hull one of eight Type 26 Global Combat

Ships, will have the same systems as HMS *London*, the last in class and which will not be commissioned until the 2030s.

The Royal Air Force faces similar challenges. The disastrous A400M Atlas C Mk 1 "tactical and strategic oversize lift" aircraft has been procured at an immense cost because part of the plane is built in Preston. Worse, the Atlas is so complex that maintenance is extremely difficult, consuming far more man hours than, say, a Hercules C-130 J, and it offers far less operational availability. Interestingly, the French and Germans are now divesting themselves of them.

Finally, the high command structure of the armed forces encourages waste. For example, the RAF has 136 air commodores, whilst the Royal Navy has a similar number of commodores. Nor is waste confined to the middle ranks. The Chief of the Defence Staff should ideally become Commander, British Forces to better forge both synergy and jointness across the forces. The only reason such a step has not been taken is that politicians and civil servants prefer institutionally weakened military leaders who pose no threat to their own power and influence. The service chiefs would then become de facto chiefs of staff for a Swedish-style Supreme Commander. As one senior officer put it "we can no longer continue to simply polish the turd."

What efforts that have been made to promote integration and innovation have too often been blocked due to little more than the politics of senior appointments. For example, a Multi-Change Integration Programme was killed for exactly that reason—it was not convenient. And yet, an examination of how to better connect capabilities through better thinking and technology is exactly what is needed, and any such programme is wholly dependent on central, high-level buy in. By developing such an innovative culture Britain could use force structures, such as the Joint Expeditionary Force, to influence the way like-minded allies and partners (and by extension NATO) do business; but all too often the services retreat into their silos and prevent such innovation. One senior officer described RAF Bomber Command at High Wycombe as a "parallel universe" and utterly resistant to joint programmes simply because there is no supreme commander to hold it to account. Worse, the MoD tends to control the services by veto rather than direction, with the driving forces being risk-averseness and the need to protect ministers from getting into trouble. Silo-building and the retreat from jointness reinforces a tendency for projects to be seen in isolation, which regularly leads to duplication of effort and cost. For example, the Future Combat Air System or Tempest programme exists only in its own bubble. What is needed is an effective Joint Requirement Oversight Committee, but for all the reasons above there is a lot of resistance to such an idea.

There are several further steps that should be taken now. The MoD should be restructured as a lean military strategic headquarters and a department of state mandated to direct and ensure coherence between all defence activities. The Chief of the Defence Staff should become the Commander, UK Armed Forces (CUKAF) and unequivocally be their strategic commander with all the authority this implies. All electronic, cyber, space and information warfare should be merged within a revamped Strategic Command which should be re-titled Joint Command. The jointness ethos at the heart of the Joint Rapid Response Force should be made central to the Joint Expeditionary Force. The latter should also be used as an instrument for NATO standardisation of equipment and procedures. Defence efforts also need to become far more integrated. Respect the histories of the three services but do not pander to them. This too often leads to a competition for resources which is the very antithesis of sound defence strategy. Reconceive professional military education to better educate middle-ranking and senior officers to think more joint and more strategic as part of a mission command culture.

Other reforms that are desperately needed include the end of management by ministerial edict which routinely puts politics in front of strategy; financial control systems that apply increased funding more effectively; and the rebuilding of a centralised relationship with industry so that critical equipment can be procured quickly. This will include the better integration of platforms with systems early in the procurement cycle. For example, the P8 Poseidon maritime patrol aircraft should have been constructed with the Stingray upgrade built in, not as an expensive afterthought. Far more kit could be bought off the shelf when necessary, especially if procurement policy is detached from industrial policy where possible. All platforms should have a full complement of weapons systems with stocks replenishment planning based on use in a high-end war. Such an approach would be aided by a technology survey designed to reinforce the connective tissue between the services by making better use of autonomous systems, such as light helicopters, Future Combat Air System (FCAS) system-of-systems technology, and the battlefield cloud. Above all, "Defence Enshitification" should be ended in procurement whereby a platform sits between partners to a contract forcing them into becoming mutual hostages of each other, with neither party being able to break out even if the value to both parties diminishes rapidly.[4]

The Utility of (British) Force

In January 2024, Foreign Secretary Lord Cameron said, "it is hard to think of a time when there has been so much danger and insecurity and instability in the world."[5] Today is not the late 1930s or late 1962, but his essential point was well-made. On 15 January 2024, on the eve of the massive NATO Exercise Steadfast Defender, London made much of committing some 40,000 British service personnel. Unfortunately, the commitment was not a coherent force like the Joint Rapid Reaction Force but what mattered was simply the politics of sending 40,000 personnel. *Plus ça change*? Three hard realities should now be apparent to London. First, the need for a British conventional force with quantity and quality that raises the threshold of nuclear use. Second, the need for a retained nuclear deterrent as the ultimate backstop against aggression. Third, sound defence is a value not a cost.

So, where would the money come from without adding significantly more to an historically high tax burden? There are still savings to be made. Back in 2014 the Taxpayers' Alliance identified some £120 billion of wasted public expenditure. The Cabinet Office Efficiency and Reform Group have identified over £50 billion of additional costs in pay and pensions for the public sector compared with the private sector, according to the Office of National Statistics. The Institute of Directors claims £25 billion has been lost because of inefficient public sector investment, whilst according to the National Fraud Authority £20 billion has been lost due to public sector fraud, whilst the taxpayer lost £4 billion alone on the sale back into the private sector of the Royal Bank of Scotland and the Northern Alliance.[6] £120 billion is over twice the annual defence budget in 2024 which would be more than enough to close Britain's defence ends, ways, and means gap.

If London does not have the wherewithal to reduce fraud there is one other saving that would afford the future force—scrap the nuclear deterrent. Both authors continue to believe that Britain should retain a credible nuclear deterrent and that an assured second-strike capability under British control enhances deterrence. However, the wilful under-funding and procurement of the conventional force is driving London inexorably towards having to make such choices unless a profound change take place.

The ends, ways and means of British security and defence policy and strategy simply no longer add up, leaving an underfunded Defence Nuclear Enterprise creaking at the seams and a conventional force constantly reducing relative operational utility. As Wyn Bowen and Geoffrey Chapman wrote in 2023,

> While the Chancellor recently committed to maintaining the defence budget at least at 2% of GDP in his November statement, a weak pound and soaring inflation would make any new procurement project difficult to keep within budget. Even with these factors in mind, it is worth considering that nuclear defence spending amounts to some 6% of the defence budget, and this arguably makes it highly cost effective compared to conventional military alternatives. But any increase to nuclear defence spending will inevitably come at the cost of the conventional side of the house at a time when the Alliance is looking to bolster conventional defences and deterrence against Russia.[7]

There is a joke doing the rounds which would be funny were it not so serious, that if the defence budget is increased much more the British Army will cease to exist. Every time the defence budget has been "increased" since 2010 the army has become smaller. The Royal Belgian armed forces are manned by excellent people several of whom are known to the authors. However, they would freely admit that the force is too small, too ill-equipped, and for too many years has had too little money invested in equipment as opposed to personnel. Not only does Belgium struggle to meet its commitments to NATO, even though NATO Headquarters is in Brussels, their forces are utterly unable to maintain interoperability with many of its more potent Allies, with little or no interoperability possible with US forces at the high-end of the conflict spectrum. Britain is heading in the same direction.

One problem is that even establishing what Britain spends on defence is not straightforward. Generally, British political discourse focuses on the "Defence Budget," but as the Ministry of Defence advised a recent House of Commons Defence Committee inquiry, "The core MoD budget is not the same thing as total government spending on Defence as reported to

NATO." Rather, the Defence Budget is the "Resource Delegated Expenditure Limit (RDEL)" and "Capital Delegated Expenditure Limit (CDEL)" set for the Department, which excludes, for example, operational expenditure.[8] In the past, the MoD did not always report consistently to the NATO International Staff using the Alliance's definitions for categorising defence expenditure.

In 2015, the MoD "updated" its approach to reporting, which gave the false impression of a boost where there was in fact none, simply because more was included in the NATO definition of defence expenditure. As Earl Howe, the then-Minister of State, told the Lords in a Written Answer in July 2015, previously the MoD had included its own bespoke budget, the cost of operations and the armed forces pension scheme. For 2015–16, it added MOD-generated income which funds defence activity, elements of cyber security spending, and parts of the Conflict, Stability and Security Fund (relating to peacekeeping, war pensions and pension payments to MOD civil servants).

The aim was to bring the defence budget in line with NATO's defence spending guidelines. The Alliance defines defence expenditure as payments made by a national government (excluding regional, local, and municipal authorities) specifically to meet the needs of its armed forces, those of allies or of the Alliance. This was as part of the commitment to a minimum of 2% GDP spending on defence annually, a target which in fact goes back to as early as 2006, although it had a low political and public profile until the NATO Wales Summit's Defence Investment Pledge in September 2014. Like-for-like comparison between the way Britain defined defence expenditure prior to 2015 and today would suggest that London spends only between 1.5% and 1.7% GDP on defence; 2.5% GDP by 2030 would amount to pretty much the same level of defence investment over this period.

Furthermore, in the mid-2000s, military activities such as peacekeeping and capacity-building were generally funded by inter-departmental common funds (later merged into the Conflict, Stability and Security Fund) rather than from the Defence budget and did not include elements reported by other countries (such as civilian pensions). After the 2014 summit pledge, those responsible for policy took a much closer interest in the reporting process and challenged this approach. The result was that Britain was able to report an increase in the Defence Budget to NATO, although London did not formally commit itself politically to meeting the 2% target until 8 July 2015.

The devil is in the detail. The Core Equipment Budget over the ten years from 2023 to 2033 is £290 billion, with the annual charge (running costs) to the defence budget of the nuclear deterrent some £3 billion (about 6% of defence expenditure). The cost of the capital programme for the new *Dreadnought*-class submarines is about £40 billion. However, there is also the cost of the new warhead to be factored in, for which a figure has yet to be published. It will cost several more billion pounds and is already suffering delay and cost overruns. This means the cost of the renewal of the nuclear deterrent equates to about 15% of that budget.

Ultimately, the British defence budget is simply too small for the force—both conventional and nuclear—it seeks to generate. Britain spent on average between £31 billion and £45 billion per year on defence (including operational expenditure) between the end of the Cold War in 1990 and 2020/21. The headline defence budget has risen marginally since then to about £50 billion per annum, but this increase has been vitiated by defence cost inflation and now equates to a cash equivalent of 2.07% of GDP. By way of context, total public (government) expenditure is about £1,100 billion per year, of which expenditure on social security and pensions is about £220 billion per year, well over four times that invested in defence. Even during the economic crisis between 1983 and 1989 Britain spent on average 4.5% GDP on defence.

In 2020, the Ministry of Defence received a four-year settlement as part of that year's Comprehensive Spending Review, with the government allocating a further £16.5 billion for 2020/21–2024/25. The 2023 Spring Budget allocated a further £5 billion and a further £2 billion for expenditure on the Defence Nuclear Enterprise up to 2027/28. However, a February 2022 Ministry of Defence report entitled "Evidence Summary: The Drivers of Cost Inflation," wrote:

> The MOD's defence inflation reports have subsequently shown that defence inflation is generally higher than measures of inflation in both consumer goods (e.g. CPI) and the whole economy (e.g. GDP deflators). For example, in 2015/16, the latest year when MOD externally published defence inflation estimates, defence inflation was 3.1% higher than the GDP Deflator.[9]

Average UK inflation between 2014 and 2023 was 2.25% with hikes in 2022 and 2023. Therefore, annual defence cost inflation has been on average at least 5% and probably closer to 8% over the same period.

The authors also have some sympathy with arguments made by another former Chief of the Defence Staff, Field Marshal Lord Michael Carver. At the time of his death in 2001 the *Guardian* wrote,

> Carver believed that nuclear weapons were for retaliation in kind only; they could not be used even to stave off imminent defeat. In that form alone, he continued to think of the deterrent as having prevented the two superpowers from going to war in and over Europe since 1945. But he constantly warned that NATO would come off worse in any nuclear exchange, whether tactical or strategic. After the fall of the Soviet Union and the Warsaw Pact, he inveighed against the vast expense and unjustified proliferation of Britain's Trident submarines. He saw "no military logic" in the decision.[10]

That there is a problem cannot be denied. As Matthew Harries writes,

> Whatever you think of [former No. 10 special advisor Dominic] Cummings, and whether you support or oppose Trident, the point is worth listening to. The UK's facilities for making and maintaining nuclear weapons, wrote Cummings in a recent post, are characterised by "rotten infrastructure" and "truly horrific bills" amounting to "many tens of billions" over the coming years. He tried to get Boris Johnson to listen to a briefing on this topic, he claims, and was told by the PM that he'd wasted his time. Even allowing for a touch of exaggeration, Cummings's basic account rings true. For decades and across Conservative and Labour-led governments, ministers and MPs have failed to pay proper attention to the detail of the UK's nuclear weapons programme. Westminster has been interested in little more than whether the parties support or oppose Trident and whether their leaders would use it. The nuclear dilemmas posed by Putin's war in Ukraine remind us that this is not good enough. And the huge sums of public money being spent on the nuclear enterprise should make this everyone's concern, whether or not they have strong views about the bomb.[11]

The conventional side of the balance is little different. In April 2021, as part of the ten-day US-led Warfighter Exercise, the 3rd Division of the British Army ran out of ammunition after eight days.

The Equipment Plan 2023–2033 written by the National Audit Office revealed a 62% hike in the budget of the Ministry of Defence's Defence Nuclear Organisation, which increased from £38.2 billion in 2016 to £99.5 billion in 2023.[12] Whilst £109.8 billion is planned to be spent on the Defence Nuclear Enterprise over the next ten years, the Ministry of Defence expects it to rise to £117.9 billion, some £7.9 billion more than expected. The Treasury is already committed to increased spending of £2 billion per annum and is reported to be willing to increase that. However, the Ministry of Defence is notorious for underestimating costs and, as ever, seeks to place too much importance on so-called "efficiencies" to save money. This time the MoD is claiming that by front-loading funding the four *Dreadnought*-class SSBN submarines will be delivered earlier than planned and thus avoid delays and cost-overruns. Experience suggests this is again the triumph of hope over experience.

The simplest solution would be to relieve the defence budget of the cost of the deterrent, but the Treasury has vetoed that. Indeed, the debate over how the deterrent should be paid for has been contentious for many years. Claire Mills and Esme Kirk-Wade in a May 2023 briefing to parliament entitled "The cost of the UK's strategic nuclear deterrent" wrote,

> In 2007 a disagreement erupted between the MOD and the Treasury over the funding of the capital costs of the replacement programme. The MOD suggested that the capital costs of procuring the nuclear deterrent had, in the past, been borne by the Treasury, a position which the Treasury refuted. The argument centred round an increase to the defence budget which was announced as part of the 2007 Comprehensive Spending Review. The CSR outlined that:
>
> The 2007 Comprehensive Spending Review builds on this investment and grows planned defence expenditure by a further 1.5% a year over the CSR07 period, rising to a total budget of £36.9 billion by 2010–11—demonstrating the government's strong commitment to defence at a time of acute operational intensity. The settlement allows the MOD to [...] make provision for the maintenance of the nuclear deterrent. As set out at the time of the Trident White Paper, provision for this will not be at the expense of the conventional capability our armed forces need. Investment in conventional capability will continue to grow over this period, as it has done since 2000.
>
> Some commentators considered this to effectively be a commitment to fund the capital costs of the replacement programme outside the core defence budget. However, when questioned on this issue by the Defence Select Committee in November 2007, the then-Permanent Secretary to the MOD, Sir Bill Jeffrey, confirmed that while additional funding had been provided to the MOD,

spending on the Trident replacement would then take place within the defence budget.[13]

The authors are not ideologically opposed to Britain remaining a nuclear power, quite the reverse. However, they are opposed to Britain wasting money on forces of little utility be they nuclear or conventional and the additional risk to those directly responsible for Britain's security and defence, the men and women of the British armed forces.

Basil Liddell Hart once said that between 1919 and 1939 the British were ostriches, and when their heads were jerked from the sand their eyes were too angrily bloodshot to keep clear sight. In that light, the Integrated Review Refresh 2023 and Defence Command Paper 2023 must be seen as tipping points for Britain's defence effort and should not be seen as simply yet more reviews. Reality needs to be gripped because London too often appears to suffer from the Dunning-Kruger Effect: British leaders lacking in competence believe that they and Britain itself are smarter and more capable than they really are, and that they are perceived as such by allies and partners. Ultimately, Britain's greatest weakness is not its inability to close its defence-strategic ends, ways and means gap, but the strategic sclerosis and illiteracy from which London suffers, and the ingrained short-termism such sclerosis fosters, allied to a determined refusal and/or inability to lead Britain to the strategic role to which a state of its power and importance could still aspire.

Furthermore, the Defence chiefs too often must plan in the absence of a settled financial framework and in the context of a Whitehall establishment that sees defence as a cost rather than a value. Military threats are emerging and changing the character of warfare, with the conditions for shock being magnified by the neglect of defenders. Given the strategic responsibilities of an advanced global trading power of some seventy million people that is also a leading member of NATO and a Permanent

Member of the UN Security Council, Britain's armed forces are simply too weak to defend either the country or its allies, and far too weak to secure Britain's critical interests. If London fails to break out of the strategic pretence in which it is trapped, it is only to be hoped that a future enemy will be obliging enough to act in such a way that Britain's defence planning assumptions do not simply collapse like the house of cards they are.

One final thought: London must begin planning now not just for 2.5% GDP expenditure on defence but 3% GDP. Nothing guarantees waste more than defence establishments spending money quickly.

7

BELGIUM WITH NUKES?

For some time, we have asset-sweated the military, compounded by a mismatch between ambition and resource that has been robustly addressed by both National Audit Office and Defence Select Committee reporting. Our strategic resilience is at risk, and we might inadvertently reduce ourselves to a smaller, static, and domestically focused land force. I am not sure that this is either the Army the nation needs, or the one that policymakers want.

General Sir Patrick Sanders, Chief of the General Staff, February 2024[1]

Strategy and Force

Whether Britain spends 2.5% GDP on defence or not, it will not be enough to meet the cost of the planned future nuclear force and future conventional force given the way Britain spends on both. Rather, both the nuclear and conventional force will continue to suffer profound tensions between ends, ways and means. Former Chief of the Defence Staff, Air Chief Marshal Lord Peach, captures the British challenge succinctly: "We must not confuse presence (Enhanced Forward Presence) and activity (Steadfast Defender) with outcomes: posture, readiness and assigned forces in line with NATO Commitments."[2] Former Deputy Supreme Allied Commander, General Sir James Everard goes further,

> Our salvation lies with and in NATO. Not spread thinly across a range of reserve and geographical tasks (as now), but by taking on a ground-holding role with forces so aligned. The benefits would not only flow from economies of scale and effort but would give the Army much needed focus and purpose to re-learn good lessons. Committed to a fight today too many of our soldiers would die for reasons that are predictable, and therefore preventable.

He continues that some will argue that British forces are

> too good to hold ground, that the role needlessly fixes the UK forever (not true), but in truth we fear that such a role would expose the true state of our nakedness ... You can play the card of UK exceptionalism when you are exceptional, not when you are not![3]

There is one overarching test of sound strategy: readiness. It has three elements. Strategic readiness concerns the ability of the state to identify and utilise all the tools available to support a grand-strategic effort, most notably war. In Britain, that no longer exists. Warfighting readiness is the ability of said state to deploy and support a force that can fight at high intensity and across several domains. It is very doubtful Britain has any such capacity or ability today. Operational readiness concerns the ability to deploy a force for a standing commitment or respond to crises. Britain would be able respond to a crisis but only if it is short, not very demanding, not very far away, small scale, and Britain has a long time to prepare.

This is not the force the Britain of today should field in the world of today. Given the relative power of Britain's contemporary economy, its place in Europe and the world, and its role in international affairs, Britain should still plan (not aspire) to field a full-spectrum, joint high-end conventional force capable of kicking down heavy doors, even in the absence of the Americans. Even if Britain could field such a force, it would still be reliant on allies with greater mass to stay and fight indefinitely. Interestingly, that is precisely the core assumption of the "Integrated Operating Concept" the British armed forces seek to deliver. It is challenging because at a bare minimum the British future force will not only need to be forged into a US-interoperable, high-end British Joint Reaction Force fully supported by C4ISR (Command, Control, Communications, Computers, Intelligence, Surveillance, and Reconnaissance) and appropriate logistics and fires, but would also need a much more balanced force with a significantly larger land force at its core.

There are several further defence planning assumptions that London needs to grip—and urgently—to demonstrate its renewed commitment to NATO. Close ties between Britain, the US and their European partners need to be further strengthened by the stated ambition to develop a British future force that would act as a leader of a permanently available NATO fire brigade for crises in and around Europe. A clear and understood British willingness and ability to act early in a military crisis by maintaining capable high-readiness "presence forces" are crucial to international confidence that Britain would reliably, promptly, and forcefully resist aggression against allies, even from fast moving, heavily armed, locally superior aggressors, and even in the face of nuclear threats.

London also needs to confirm the shift away from a counter-insurgency and stabilisation focus to defence and deterrence to ease pressure on US forces and by being at the core of a new multinational command hub. This would enable Britain to act as a framework power for increasingly capable European coalitions, operating within or, if necessary, beyond Alliance territory, not unlike the JEF. Implicit in such a goal is full commitment to NATO's European pillar with a determination to establish a clear leadership role in it, preferably by building on the new Allied Reaction Force to augment and reinforce deterrence by having the proven capacity to prevail in intense conflict scenarios. Such a force will also need to be of sufficient mass to reinforce front-line states on the eastern and southern peripheries of the Alliance under Article 3 of the North Atlantic Treaty. Command arrangements would also need to be flexible and possibly include new arrangements with France and Germany, possibly under the auspices of President Macron's European Intervention Initiative.

Britain also needs to lead the modernisation of Alliance forces but take a pragmatic national approach to emerging and disruptive technologies by purchasing off the shelf where possible. Britain should also seek to act as a project leader of lean and mean development programmes with allies and partners particularly where such technologies offer a clear comparative advantage. London also needs to further adapt whole force doctrine, training, and professional military education in the Alliance.

London is well-placed to create a genuine strategic public-private defence and technological industrial partnership in which risk, innovation, cost, and investment is shared across the Alliance. There will be no magic high-tech money pit to nourish national industrial champions as export earners from a few exquisite systems. Britain should also be at the forefront of the information war through the refinement and integration of military and criminal national intelligence resources. It needs to re-cultivate the diplomatic skills to bring influence to bear over allies and partners in order to detect, acknowledge and overcome political warfare, covert influence, and non-lethal attacks.

The Future Force Britain Will Get

Royal Navy

By 2030, the Royal Navy Surface Fleet will comprise: two *Queen Elizabeth*-Class aircraft carriers, six Type 45 destroyers, eight

Type 26 frigates, five Type 31 frigates, and two assault ships. There will be seven *Astute*-class nuclear powered attack submarines and four *Vanguard*-class nuclear powered ballistic missile submarines with their replacement *Dreadnought*-class under construction. The Royal Fleet Auxiliary operates three *Bay*-class dock landing ships plus seven replenishment-at-sea ships and a multi-role ocean surveillance ship designed to protect critical underwater national infrastructure such as pipelines.

The Royal Navy is investing £40m in the Royal Marines Future Commando Force, whilst an additional £50m is being spent on converting a *Bay*-class support ship into a littoral strike ship, although there have been questions asked about the future of HMS *Bulwark* and HMS *Albion*. Two littoral response groups have also been stood up and have been deployed first to the Euro-Atlantic Area of Operations in 2021, and then to the Indo-Pacific in 2023. Surprisingly, given the Royal Navy's undoubted expertise in countermine operations, all mine countermeasures vessels will be retired. An automated mine-hunting capability has been developed in partnership with France and is being brought into service crewed by civilian auxiliaries. It is questionable whether it will offer the same level of capability for which the Royal Navy is renowned, even within the US Navy!

Several offshore patrol vessels will be permanently deployed to the Falklands, the Caribbean, Gibraltar, and the Gulf to enable Royal Navy frigates and destroyers to be used more efficiently and effectively. Type 26 and Type 31 frigates will be slowly brought into service and will provide protection for the littoral response groups to conduct strikes from the sea in support of resilient ship-to-objective manoeuvre. The air defence Type 45 destroyers are to be upgraded whilst the Harpoon anti-ship missile system is phased out. A concept and assessment phase will also be commissioned for a new "Type 83 Destroyer" designed to replace the Type 45 by the late 2030s, whilst three new Fleet

Solid Support Ships will be constructed around a Multi-Role Ocean Surveillance capability, plus two Multi-Role Support Ships, by the early 2030s.

British Army

Lieutenant-General Sir Roland "Roly" Walker, the Chief of the General Staff, said that his aim is to concentrate on ensuring Britain's "mid-sized army" is fully integrated with the other services and allies. The problem is that the British Army is not mid-sized, it is small. Worse, the army will be reduced from a "full-time trade-trained" strength of 76,000 today to 72,500 by 2025 and re-organised around two Heavy Brigade Combat Teams. The Future Soldier concept champions enhanced firepower, protection, and mobility. To that end, the future force will be equipped with the multi-mode Boxer armoured vehicle, the Ajax armoured fighting vehicle, and the upgraded Challenger 3 main battle tank. As well as the two Heavy Brigade Combat Teams there will be two Light Brigade Combat Teams, one Deep Recce Strike Brigade Combat Team, an Air Manoeuvre Brigade Combat Team, together with a Combat Aviation Brigade Combat Team.

148 Challenger 2 main battle tanks are to be upgraded (and called Challenger 3) under a £1.3bn modernisation programme, whilst the rest will be scrapped. All Warrior AIFVs (armoured infantry fighting vehicles) will be retired and replaced by the up-armed Boxer MIV (mechanised infantry vehicle) currently under construction at a plant near Stockport, with the programme due for completion in 2025. The plan to procure Ajax fighting vehicles remains confirmed for the moment despite the procurement challenges the programme is facing.

A new Ranger Regiment was stood up in August 2020 and forms the core of a new Army Special Operations Brigade, which draws its personnel from Specialised Infantry Battalions. £120m is also being invested in the Ranger force over the next four years. A Security Force Assistance Brigade is also being stood up to be deployed regularly at short notice world-wide in support of British defence outreach. This brigade will work with partners to train, advise, enable, and better integrate their units at the tactical level (division-level and below). It will also build relationships with key security organisations in partners and, in time of war, enable them to be integrated effectively into UK and NATO forces. The brigade has four regular battalions, a reserve battalion, and an outreach unit which deploys globally.

£250m is also being invested by 2030 in longer-range artillery fires such as the Guided Multiple Launch Rocket System (GMLRS) and £800m will be invested over ten years in a new mobile fires platform. The Exactor missile system will be retained and upgraded for longer-term use. Some older Chinook helicopters will be retired from the Joint Helicopter Command, and newer extended range variants purchased, along with a new medium-lift helicopter by 2025 as part of a consolidation of the army's fleet. The army will also upgrade Watchkeeper unmanned aerial vehicles and improve air defence against both aircraft and drone attacks. Critically, the army needs far more drones, both in reconnaissance and strike roles, and to up its game in electronic warfare.

Royal Air Force

The RAF will retire its Tranche 1 Eurofighter Typhoons and Hawk T1 trainers/Red Team aircraft by 2025, whilst the C130J Hercules fleet was retired in 2023 and replaced by the Airbus A400M. This increases maintenance costs significantly and raises questions about suitability for some missions and fleet sustainability. The Typhoon force will also be upgraded with new weapons systems, such as the SPEAR Cap 3 air-launched precision guided missile and the Radar 2 programme. Strategic lift has been maintained with retention of the RAF's eight Globemaster C-17 aircraft and a major capability gap filled with the procurement of six Poseidon maritime patrol aircraft.

£2bn is also being invested in the Tempest Future Combat Air System over the next four years, whilst the E3D Sentry was retired in 2021 to be replaced by three new E-7A Wedgetails that have been delayed and are yet to enter service. The current order of forty-eight F-35B Lightning 2s will be increased to seventy-five, which is still markedly lower than the 138 aircraft originally planned. Britain will also invest in the software and weapons upgrades needed to ensure the F-35 force has the constant data interface upgrades which are the *sine qua non* of the aircraft and which could, in time, turn it into a force hub for an array of deployed "loyal wingmen" drones. This is important given that by 2030 the RAF believes 80% of air operations will be unmanned.

Updated Defence Planning Assumptions

Britain, like any state, should base its defence on hard facts and evidence. It is a heavily populated island off Northwest Europe on the great circle route between Europe and the United States. Britain also dominates the exits from the Baltic and the Arctic into the Atlantic. It must trade or its population starves and freezes. Britain also has limited industrial capacity and is dependent on the supply of raw materials, with a very limited defence and technological industrial base whose forces and resources are concentrated in a few places. The systemic interdependence of its communications, supply chains, and resources makes the British way of life acutely vulnerable, particularly to cyber-attack. That is what London must defend.

The defence of the kingdom rests first and foremost on the maritime control of the North Atlantic and ensuring and assuring rapid American and Canadian reinforcement of Europe during a pre-war emergency. This is also a vital part of NATO's deterrence posture. Britain is also vulnerable to air and missile attack, as well as a blockade. The threat of air and missile attack stretches in an arc from Greenland and the North Cape to beyond the River Oder, the north coast and the Mediterranean and to the Azores.

Britain is a nuclear power with a sea-based nuclear deterrent, but dependent on the Americans. The Royal Navy has some limited carrier strike and anti-submarine capability but little else. None of the armed forces have reserves to speak of. In 2010, it was hoped Army reserves would make up the numbers, whilst the Royal Navy and the Royal Air Force effectively have none.

Britain lacks political leadership and direction and is becoming an increasingly insular and divided society ethnically and socially, with an economy increasingly and overly dependent on services and not manufacturing. The future direction of the economy is uncertain, whilst domestic infrastructure is underfunded and vulnerable to attack with an ageing population being paid for by taxing the young. Unless there is a profound change in the way London thinks and acts the level of risk faced by the British people could rapidly become extreme.

Britain remains vital as a strategic base for the Americans and NATO, while control of the airspace over Northwest Europe is vital to Britain's defence. In other words, Britain needs NATO and NATO needs Britain. However, to control the northern littoral of Europe, which is vital to prevent blockade and prevent invasion of parts of the continent vital to Britain's defence, London must retain the military capacity and capability to enter the rest of Europe and defeat an enemy as part of a credibly effective NATO alliance.

In the worst case, British forces must also be able to enter Europe from more than one direction during an emergency, not just across the Channel. However, lacking reserves, Britain cannot afford to get into a long war, which is exactly what Russia would probably aim to do. Therefore, at the very least, the home base must remain secure at the greatest distance possible from the enemy to allow time to build up forces and resources. The priorities for the allocation of resources will be the military defeat or deterrence of a patent threat; effective diplomatic, informational, military, and economic tools of power; and to prevent a latent threat becoming a patent threat in such a way that, if all else fails, the nature of the threat is identified in enough time and distance to mount an effective defence. Russia, by its actions, has become a patent threat to at least the eastern nations of NATO, and as an ally the UK must act accordingly. Therefore, at a time of scarce resources priority for allocation of said resources must be where British forces can best add value and in this order: 1. the strategic home base, 2. Northwest Europe and the North Atlantic, 3. Eastern Europe, 4. Southern Europe and the Mediterranean.

Britain must also decide if it wants to have very significant influence over NATO operations and fight its own battles, or be a bit part player in US-led operations. Britain will only gain significant influence if it makes a very significant contribution to NATO's collective defence. At the very least, London must assume the need for very significant influence over the defence of the strategic base with sufficient deployable forces to fight its own battle if needs be in Northwest Europe, even if as part of an Allied force.

What Will NATO Need?

What will NATO need? In some important respects, Article 3 of the North Atlantic Treaty is just as important as Article 5 and a test for whether Britain is serious about placing itself at the heart of NATO. Article 3 states that "separately and jointly, by means of continuous and effective self-help and mutual aid, [allies] will maintain and develop their individual and collective capacity to resist armed attack." Against the backdrop of increasing US military overstretch Washington is likely to demand NATO Europeans provide collectively by 2030 at least 50% of all designated capabilities. The NATO Defence Planning Process also has a cardinal rule that no single ally should provide more than 50% of any agreed Capability Target. Compliance with that rule has failed in far too many "strategic enabler" domains, with the United States providing 70, 80 or even 90% in various categories. Over the longer term (by 2035), NATO Europeans will probably have to deliver two thirds (67%) of NATO's combined operational capacity for collective defence, as measured in rapidly usable forces, enablers, and other capabilities across SACEUR's Area of Responsibility.

The New Force Model is at the heart of NATO's current planning and Britain has placed NATO at the heart of its defence strategy. The NATO Military Strategy 2019 calls for the Allied Reaction Force of some 40,000 troops to be transformed into a future force of some 300,000 troops maintained at high alert, with 44,000 kept at high readiness. At American behest the new force will be mainly European. A force of that size and with the necessary level of fighting power would normally mean that with rotation there would always be a force of some 100,000 ready for deployment at relatively short notice, which will be extremely expensive for NATO European allies grappling with high inflation and post-COVID economies. A NATO standard brigade is normally between 3,200 and 5,500 strong (a British brigade is about 2,500 strong). Given that both air and naval forces will also be needed as part of the joint force concept, a land force of, say, 200,000 would need at least fifty-to-sixty European rapid reaction brigades together with all their supporting elements. At best, there are currently only twenty-to-thirty.

Therefore, if the Alliance's "family of plans" is to be credible going forward NATO Europeans, with Britain at their core, will need to take on far greater strategic responsibility. The good news is that since March 2024 Finland and Sweden are now NATO members and bring significant additional military capability and capacity to the Alliance, even if they also bring significantly expanded commitments. NATO will also require a profound change in the culture, capabilities and capacities of British forces, their combat support and combat support services, but, above all, in the political mindset of leaders.

This is because reinforcing NATO's conventional deterrent will be a dynamic process of adaptation that never ends. Today's NATO Force Model is driving the development of the two key elements of Allied force modernisation: the enhanced NATO Response Force and the Allied Reaction Force. However, in time both will need to become heavier and very mobile forces able to act across multiple domains as part of new Warfighting Corps. Such modernisation in both the strength and weight of force will also require additional investment in capabilities and capacity by European allies. Only the Alliance can provide the necessary balance between capability, capacity and affordability that will need to be struck through greater efficiency and enhanced effectiveness.

Even to fulfil the commitments to increased defence investment and force levels made at the Madrid, Vilnius and Washington summits will require all NATO Europeans to significantly invest more in defence if they are serious about a strong European pillar inside a credibly strong NATO. For London, the challenge is implicit but clear: Britain must become a command core of a European-led high-end, first responder force in an era in which NATO must not only plan for territorial aggression in Europe but also assume it will likely coincide with aggression elsewhere in the world that will demand US attention and forces. It is questionable whether US forces are any longer sufficiently capable of meeting several high-end contingencies simultaneously, which is why the allies are important to them, and why future NATO military strategy must assume that European members of NATO will be able to defend Europe at any time and in any circumstances when the bulk of US forces could be engaged globally.

Given this requirement, what will NATO want from Britain? If European allies collectively assume a greater share of the responsibility over time to assure, deter and defend the Alliance Britain will need to be in the vanguard along with France and Germany. Allies must also adhere to four principles which relate to the overall operational capacity required, the individual national capabilities that must underpin that capacity, and the necessary supporting defence investment effort.

Principle One. By 2035 European allies must provide collectively two thirds or more of NATO's overall required operational capacity, as measured in rapidly usable forces, enablers and other capabilities needed to execute advance plans across SACEUR's Area of Responsibility.

Principle Two. No ally will be expected to contribute more than 50% of any individual NATO capability area, as pursued through the NATO Defence Planning Process. European allies must thus aim at providing collectively two thirds or more of any given capability area, recognizing that progress will be easier and faster in some areas than in others.

Principle Three will be the hardest but most necessary. By 2030, European allies must invest at least 3% of their Gross Domestic Product (GDP) annually on defence to remedy existing shortfalls and meet the requirements across all domains arising from a more contested security order. They must also plan for further increases in their respective defence budgets so that in a pre-war emergency they can invest additional moneys effectively and expeditiously. Allies must also continue to invest at least 20% of their defence budgets on major equipment, including related research and development.

Principle Four. NATO Europe's combined operational capacity must aim to provide at least four fully capable, fully enabled, fully ready Warfighting Corps. These four corps would be sourced from the eight Rapid Reaction Corps Headquarters in the current NATO force structure and available on short notice in times of tension or war. Each of the four corps will also require contributing nations to source all the required combat, combat support and combat service support units.

In time, a European-led division-strength, air, sea, and land force that takes the NATO Response Force to a new level of capability will be needed if the central plank of NATO strategy, deterrence by denial, is to remain credible and the threshold for the use of nuclear weapons in Europe kept necessarily high. Such European capability will also be demonstrable burden-sharing by easing pressure on US forces and resources and by providing a high-end focal point for the transformative NATO New Force Model and future readiness initiatives.

Unfortunately, the NATO Response Force, an otherwise brilliant concept, has had an unintended consequence in that the rotational leadership principle encourages a fragmentation of capabilities among European allies. While the United States has been building up the US Army's V Corps stationed in Germany and Poland, European allies have exhausted themselves trying to build up ten army corps concurrently. And, whereas the US Army's V Corps has most of the enablers it would need to fight a high intensity war, none of the European corps do, with the partial exception of the UK-led Allied Rapid Reaction Corps (ARRC).

One option London could consider would be to redeploy HQ ARRC from the UK to Poland to become the headquarters of the new Allied Reaction Force (ARF). To do this London would need to transform HQ ARRC from being a high-end warfighting corps into a sophisticated and demanding deployable Joint Force Headquarters. Equally, such a commitment would reinforce HQ Multinational Division (Northeast) and enable British forces to better work alongside Polish forces and US V Corps to create an "unblinking eye" on NATO's east. This British-led force would also be able to move rapidly anywhere in SACEUR's Area of Responsibility and it could also act as a rapid reaction force able to move rapidly from a low-end to a high-end conflict across the full spectrum of missions, as well as a rapid deployable strategic reserve. HQ ARF would also reinforce deterrence simply with its presence in Central Europe rather than Western England. Moreover, because it would have a focus on NATO's northern, eastern, and south-eastern flanks it could also offer support to allies on the southern flank if called upon. Whilst a considerable challenge, the ARF should aspire to become a high-end, first responder Allied Future Force able to act from seabed to space and across the multi-domains of air, sea, land, cyber, space, information, and knowledge. The ARF must also be sufficiently robust and responsive, and held at a sufficient level of readiness, to meet all threats to the territory of the Euro-Atlantic Area in the first instance and have sufficient capacity to support those frontline nations facing transnational threats, such as terrorism.

There are a host of other critical NATO capabilities in which Britain should invest: deep fires, battlefield air defence, combat engineers, wet gap bridging, movement control, medical support, ISTAR (intelligence, surveillance, target acquisition, and reconnaissance), and electronic warfare. Similarly, instead of having six half-empty Joint Force Air Components, European Allies must commit to building three, fully structured, fully capable Composite Air Strike Forces (CASFs), each with the full complement of fighters, fighter-bombers, tactical reconnaissance, electronic combat, airlift, tanker, early warning and combat search and rescue aircraft. These three multinational CASFs would work together with US European Command (USEUCOM) 3rd Air Force, thereby providing the SACEUR with four potent, full spectrum air packages. The same template must also apply to navies with the activation of European Standing Fleets in the Atlantic and Mediterranean operating alongside and together with the US Navy's 2nd and 6th Fleets. This will include NATO cross-attachment of European surface ships, submarines, amphibious forces, and maritime patrol aircraft to US Navy naval task groups and similar US Navy assets being cross-attached to European naval task groups.

NATO Integrated Air and Missile Defence must be extended to allow earlier intercepts, including "left of launch," to better protect deployed forces. The war in Ukraine has shown that however capable are systems such as Patriot, SAMP-T, and others, relying on point defences alone will not protect critical permanent (non-mobile) headquarters, facilities, and assets when confronted with swarms of incoming ballistic and cruise missiles and drones. NATO's premier Aegis-based ballistic missile defence system, currently focused on Iranian missile threats, must also be optimized to create a "360 degree" capability against Russian ballistic and cruise missiles. This would require not only a policy decision, but also deployment of a TPY-2 advanced mobile radar in Poland or Bulgaria.

In addition to the 50% rule, NATO Europeans must also commit to increased investment across the three core missions of collective defence, crisis management and co-operative security. Britain has just about met the current minimum NATO baseline of spending at least 2% of GDP on defence and a real terms increase to 2.5% by 2030 is to be welcomed, but for all its financial travails, if deemed sufficiently important Britain also has the capacity to do more, such as including measures that enhance security in addition to defence spending, with systems in place that would enable London to increase both investment and outputs rapidly in an emergency. Ideally, Britain should aim to spend at least 6% GDP on enhancing security in all its forms, of which at least 3% should be on defence, and no later than 2030. This is important not just for Britain because some of the smaller allies will only make such a move if countries like Britain lead the way.

There is another more intangible factor. Given history, both Europeans and North Americans expect Britain to play a leading role in the defence of the continent. This is because Britain still has a residual strategic brand that requires London to be at the forefront of *all* NATO defence efforts. Power is as power does and London cannot hide from Britain's own power, much though it tries to. The NATO bottom-line is this: in a major pre-war emergency two fully capable, fully enabled, fully ready "shield" corps (Multi-National Corps, Northeast in Poland, and Multi-National Corps, Southeast in Romania) will be needed, supported by the Allied Reaction Force. They will constitute NATO's first line of land defence and play a key role in protecting forward-located allies, delivering on the NATO commitment to which Britain is signed up not to yield Allied territory, and buy time and manoeuvre space for the four Warfighting Corps. NATO Europe, with the support of the Americans, will need to strengthen the operational capacity and credibility of these two multinational corps, including the provision of combat support and combat service support units. Only then can the current posture of deterrence by denial be credible.

European Composite Air Strike Forces will also need to be reinforced by two fully capable, fully enabled, fully ready standing fleets in the Atlantic and Mediterranean. The two fleets would be sourced from the current six European Maritime Force

(MARFOR) Headquarters, with the naval task groups always having a minimum operational capacity that can be augmented by US forces at short notice. The core of the two fleets would be the aircraft and helicopter carriers operated by France, Italy, Spain, but above all the two *Queen Elizabeth*-class ships of the Royal Navy, together with their NATO European complement of surface escorts (Anti-Air Warfare destroyers and Anti-Submarine Warfare frigates), attack submarines and maritime patrol aircraft. Such a commitment would also require a marked increase in British commitment and investment, as the two carriers are not currently envisaged to be available at the same time.

Such a commitment would not only give Britain the leading role in NATO to which it claims to aspire, it would also preempt growing American dissatisfaction with the burdens the Alliance imposes on its own forces. For its part, the United States would commit to permanently stationing in Europe, under the command of the USEUCOM, a fully capable, fully enabled and fully ready warfighting corps (US Army's V Corps); a fully capable, fully enabled and fully ready CASF (US Air Force's 3rd Air Force); and a fully capable, fully enabled and fully ready US Navy 6th Fleet and its NATO component (STRIKFORNATO) for Allied multi-carrier operations, and complemented by US Marine Corps and Special Operations Forces. These forces would meet the same NATO standards as their European counterparts to ensure critical interoperability at the high end of warfighting when both command and forces are under intense pressure.

Once More Unto the Breach?

What could Britain offer? Britain is unlikely to be able to fund all the forces and resources NATO would ideally like London to afford. Indeed, a central theme of this book has been the hard choices Britain will need to make, even if London did spend 3%

GDP on defence. The bottom-line of the future NATO Britain force is this: if London cannot put a heavy division in the field and keep it there for at least six months, it is offering little more than Canada, and London will lose influence to others in both Washington and Brussels. Therefore, for Britain to properly meet its minimum NATO capability requirement it would need to increase the size and enhance the capability of the British armed forces by roughly 10%-to-15% from warfighting teeth to combat support tail. A defence budget of approximately £50 billion or 2.08% GDP is simply insufficient to enable London to play the leading role in the Alliance to which it claims to aspire. This is not least because of the way Britain invests in defence, the costly and timely procurement process, and because the cost of the nuclear deterrent is borne by the defence budget and not the national contingency reserve. If efficiently spent, a defence budget of 3% GDP or £90 billion could be enough.

The numbers the British armed forces require are difficult to judge without a concept of operations which dictates the organisation. Big groupings should in any case be avoided as much as possible as they and their logistic support are vulnerable to discovery and destruction, so the shift to small groups with their efforts focused rather than massed is probably the way to go. However, by way of example, Operation Granby during the 1991 Gulf War involved about 20,000 British personnel, 7,000 vehicles and seventy aircraft. To sustain these forces in battle for any duration requires the numbers to be trebled at a minimum. Therefore, to meet those commitments at the very minimum the British armed forces would need to be roughly the same size as they were in 2000 when the British Army had 110,000 regulars and 45,000 reserves, the Royal Navy was 45,000–50,000 strong, with some 25,000 reserves, and the Royal Air Force had about 40,000 regulars augmented by a highly skilled reserve force of some 20,000. According to Statista, as of 24 January 2023 the number of personnel in the British armed forces was 142,560 regulars and reserves.[4] Britain needs at least 205,000 regulars and 90,000 reserves by 2035 at the very latest.

Royal Navy

The Royal Navy should focus on developing two force drivers: delivering the deterrent and protecting sea lines of communication in the Atlantic, North Sea and approaches to the Baltic. A continually-at-sea nuclear deterrent must be maintained but with enhanced security and able to operate securely in "de-sensitized" areas. The Royal Navy should provide the core component of a new non-American carrier-strike capability using existing British, French, Italian and Spanish aircraft carriers, with the Royal Navy's Carrier Strike Group protected by Belgian, Dutch, Danish, Norwegian, and other allies. The Navy also needs to ensure it has the capacity to deliver a sufficiently wide-area anti-submarine capability to properly protect the nuclear deterrent. A full upgrade and upgunning of the six Type 45 destroyers is needed to provide extended area defence and they must be armed with new systems, such as long-range PAAMs missiles.

The SSN-AUKUS programme of planned nuclear-powered fleet submarines (SSN) should be exploited to increase the number of Royal Navy SSNs to a minimum of ten with, at any one time, a deployed NATO Northern Atlantic Fleet protected by at least three SSNs. A new arrangement is also needed that would afford the navy greater control of dockyards. The Royal Navy focus must be on a credible and effective role securing sea lines of communication where it can add most value to the US Navy and NATO allies, in the North Atlantic and in the North Sea and Baltic approaches. That means by 2035 at the latest a minimum of twenty-four Type 26 and Type 31 frigates, six-to-eight Type 45 or Type 83 guided missile destroyers, four nuclear powered ballistic missile submarines, ten nuclear-powered attack submarines, plus supporting amphibious assault ships, minesweepers, logistics vessels and patrol craft.

All necessary ships should be procured to sustain the fleet on operations and reconceive littoral strike to ensure it does not drain resources from the nuclear deterrent and carrier strike missions. Finally, it should be ensured that flag officers serve longer in their assigned positions, logistic support meets the requirement for NATO exercises and operations, and maritime air capability is sufficient for the roles and tasks expected of it out to 2040. The Merlin helicopter force is being "sweated too hard" at present.

British Army

The army only keeps half a brigade at readiness but to meet the NATO Defence Requirement it should be able to deploy two heavy warfighting divisions each with a minimum of two armoured regiments. Such a force is far beyond the army's current capability. Army Command is also still structured as if it was a large force when in fact it is very small by international standards.

Furthermore, the future British Army commitment to NATO is to supply eight army brigades as part of the 2024 and 2025 iterations of the NATO Defence Planning Process. The current plan is for the entire Army to be only eight brigades strong, but force rotation would require three times that number. In other words, there is no possibility that the British Army could provide such a force of regular personnel and at graduated readiness: at least three of the formations will be held at low readiness and comprised of reserves. London is thus "playing innovative tunes" over the use of "reservists" so that on paper at least Britain can offer NATO two deployable divisions—one "armoured" and one "mechanised," albeit with their frontages doubled from around 10 km to 20 km. Even to realise such a force the British Army will need more armour and longer-range fires than planned, and will rely on emerging and disruptive technologies, such as AI-enabled drones to offset the dramatic reduction in infantry numbers and fighting power. The British Army will also need much improved command and control, war stocks and, above all, logistics.

The future army needs at least one corps headquarters (HQARRC) with all necessary enablers, including a corps "fires" brigade. A minimum of one warfighting division able to enter battle anywhere in Europe in thirty days or less is needed, with another ready to fight in ninety days, also with all enablers, most notably artillery, both tubed and rocket, as well as engineers and army aviation, with adequate stocks of spare parts and ammunition. UK Special Forces, 16 Air Assault Brigade and the Royal Marines must also further develop their capacity to undertake very high-end early-entry and out-of-area missions, albeit of limited scale and duration.

Whilst all of Britain's armed services have had to weather cuts since the turn of the twenty-first century, the army has faced the most turbulence in its core purpose. After the comparative stability of the Cold War, ready to fight the Soviets as the British Army of the Rhine, the army's focus has been buffeted by frequent shifts of emphasis in UK foreign policy. This started with international policing in the 1990s as a "force for good" under UN auspices, followed by a sustained best effort on counterinsurgency campaigns in Iraq and Afghanistan, then a period stressing global engagement and "grey zone" competition, and now back to conventional deterrence and defence in the Euro-Atlantic region as part of NATO. This discontinuity has taken its toll on the army's capabilities, delaying by at least a decade the modernisation needed for peer warfighting. The net result is the UK being adrift of its modest NATO targets on land, both quantitatively and qualitatively.

However, green shoots are emerging from a renewed UK Defence focus on leadership within NATO, attendant with a stronger continental commitment. This promises to deliver correspondingly capable land forces. Whilst the 2021 Integrated Review can seem questionable in hindsight, not least by cutting the size of the regular army by nearly 10,000 less than a year before a major land-based threat materialised in Europe, it did allocate circa £25bn to new equipment. As a result, the army is finally on the cusp of being recapitalised. Over 500 Ajax reconnaissance platforms will be fielded; 623 wheeled Boxer armoured vehicles are on contract and being built in the UK; and 148 Challenger 2 tanks will be upgraded to Challenger 3. This is an historically low number for the UK, but Challenger 3 promises to be one of the best tanks in the world, and contemporary battlefields are changing emphasis. Long-range capabilities look ascendant. Hence, a vastly more capable attack helicopter (Apache 64E) is now being introduced; seventy-two multiple launch rocket systems are being upgraded, able to fire a variety of advanced munitions; and a modern 155mm artillery system will be arriving by the end of the decade to complement the Swedish Archer howitzer already in service. Taken together with uncrewed systems, from small lethal drones up to Watchkeeper with its wide-area radar sensor, this will revolutionise the way the army fights—linking reconnaissance to strike in ever faster cycles, deeper into enemy rear areas.

The aim is to offer NATO a competitive warfighting corps, with two small but still—just—militarily effective divisions. One of these will be focused on armoured fighting vehicles. Albeit slower to deploy, it still offers the hardest offensive punch and could operate as the nucleus of NATO's strategic reserve corps. The other division will be lighter, aimed at responding quickly to deter or counter aggression. Its lethality will stem from autonomous systems, longer-range anti-tank weapons and combat aviation. Army special operations forces and security force assistance expertise, in concert with 77th Brigade information operations, completes the full breadth of the offer. If these initiatives are delivered comprehensively, Britain may once again be proud to possess the small but highly capable army NATO needs, referenced throughout the world. But it is a big *if*—dependent on the Ministry of Defence staying the course on recapitalisation and investing in uncrewed technologies.

Royal Air Force

The Royal Air Force needs to get back into the business of long-range strategic effect, reinforced by a restored power projection mindset. In recent years it has become too tactically focused on the air-land battle. Looking to the future London must give the RAF the capacity for much heavier strategic punch by equipping it with aircraft such as the B-21 Raider. More also needs to be done by RAF commanders to better understand how best to control the air space in a joint environment.

The RAF should consider now whether control of the air needs expensive manned systems, such as the E7a Wedgetail, or if such platforms could be supported and/or replaced by long-duration drone systems. A better understanding of the role of the army and its needs for strategic and tactical lift is a priority. In the short-term all crude systems and infrastructure, such as the Wildcat missile system, should be discarded. These systems and facilities require too much commitment for too little effect.

Above all, it should be ensured that a minimum fleet of seventy-five F-35 strike aircraft is procured through working with the Americans to enhance and improve the capability. It should also be considered how Coastal Command might make better use of autonomous systems such as Wave Glider drones, simplifying the "spaghetti command" of the small British helicopter force, placing all of them under RAF command and give the RAF responsibility for control and exploitation of the airspace and the electro-magnetic spectrum. Some choices will hurt. For example, the RAF Red Arrows aerobatic team places too much of a burden on frontline squadrons, costing 1.5 pilots for each of the ten frontline combat squadrons. This must end. Frontline pilots should also be placed at the forefront of career progression to improve morale. In 2023, several F-35 pilots resigned. They are too valuable a resource to lose.

The Royal Air Force currently has some 160 air defence fighters and ground attack fighters in its inventory, which at a steady state operational readiness rate of 75% yields 120 operational aircraft at any one time, which is about the size of the Ukrainian air force. A major NATO operation (for example, to defend the Baltic States) could conceivably include British, French, Nordic, and Polish air wings of fifty or so combat aircraft each, plus three wings of US Air Force and a composite wing made up of smaller allies. Other critical requirements are the requisite number of tankers; intelligence, surveillance and reconnaissance (ISR) assets; and maritime patrol aircraft to support the RAF's committed NATO roles and missions.

There are other actions that the RAF should consider. Where possible buy ISR assets commercial off the shelf. There are also still too many bespoke combat ISTAR systems. The RAF also needs to think more carefully about its role in defending critical undersea infrastructures. The number of available pilots is only just enough for the "peacetime matrix" and needs to be increased because in an emergency the RAF would very quickly run out of pilots. To that end, RAF reserves need to be markedly strengthened and given core roles, such as flying aircraft as part of the Air Defence of Great Britain, much like the US Air National Guard. This would release regular officers for deployed NATO duty.

Much more use needs to be made of advanced simulation and the synthetic environment in training, allied to much improved mission-command-led professional military education. Given the availability of simulators, the "White Scarf Brigade" mentality needs to be stopped. The Typhoon force of 160 aircraft should give the RAF about 1 million flying hours but many of them are wasted through unnecessary sorties by pilots which could be carried out on the simulator.

Finally, and not least, the RAF needs to make far better use of its highly skilled ground crews and their expertise. A Shadow Personnel Plan should thus be established to attract highly skilled members of the civilian workforce on a pro tempore basis.

Belgium With Nukes?

Britain is not Belgium with nukes and the reasons a country like Britain invests in defence do not simply concern national defence. A relatively strong British defence effort buys influence inside NATO, with partners, in the UN and G7, and above all in Washington. Even 3% of Britain's GDP spent on defence would be barely enough, given the strategic circumstances which the British must confront, but it would be an advance. Given the year-on-year increases in defence expenditure by China, Russia, Iran, North Korea, and others, 3% GDP on defence could only ever be enough if Britain spent far, far better.

In that light all the recent reviews are little more than down-payments on sound defence to alleviate the eternal short-term funding crisis from which Britain's armed forces suffer, with AUKUS a further down-payment towards a new intelligence-led Five Eyes-based alliance fit for the twenty-first century. The problem is that the first is little more than a fix and the second reflects a strategic shift to a Global Britain that in defence terms is at best a distraction from NATO. AUKUS might in time reduce the unit cost of SSN (debatable) and promote a more joined-up security, defence, and industrial policy (questionable), but the Indo-Pacific is not the centre of gravity of British strategy, be it grand, national, or defence, and never will be. Europe is. Experience would instead suggest that AUKUS is going to prove very expensive, particularly if the Americans cut the number of Virginia class SSN on offer to the Australians, and possibly beyond Britain's industrial capacity to deliver. In other words, whilst AUKUS implies a Global Britain view of Britain's role in the world and the utility of defence therein, the recent defence reviews do the opposite and clearly limit the very defence investment central to the AUKUS vision. The result? A grand strategy that bears little or no relation to defence strategy and which not for the first time implies ever more tasks over ever greater distance for Britain's dangerously small, hollowed-out and overstretched armed forces.

The hard truth is that in relative terms the British armed forces are too small and too ill-equipped for the missions and tasks the British government has signed up to in NATO let alone Global Britain. Many of the commitments made, most notably the Future Soldier, also remain unfunded. Consequently, as currently envisaged the British future force have neither the quality nor the quantity to meet existing let alone future NATO commitments. This is particularly the case for the British Army which has become a leitmotif for the dead-end which for too long British defence policy has been parked in. It is precisely for that reason the British Army of today faces a very similar challenge to Lord Gort's force back in 1940. For all its illustrious history the British Army of 2024 is an "anything-but-warfighting" force that might soon find itself fighting forces that exist only for warfighting. Much the same can be said for the Royal Navy and the Royal Air Force. Leo Tolstoy wrote in *War and Peace* that, "The strongest of all warriors are these two—Time and Patience." Some real, relative, and relevant firepower also comes in handy.

8

TOMMY

Advance Britannia!

Winston Churchill, May 1945

That Which We Are

The Israelis have a saying, "weakness is a provocation." In May 1945, Aneurin Bevan, a British politician, lamented, "This Island is made mainly of coal and surrounded by fish. Only an organizing genius could produce a shortage of coal and fish at the same time."[1] In 2022, Britain faced a drought, an energy crisis, and the price of fish went through the roof. Britain is not alone. Germany is facing an energy crisis entirely of its own making, whilst France faces extreme drought and an energy crisis due to faults in its ageing fleet of nuclear power stations. The deeply ingrained cause of their respective crises is not simply the unexpected and appalling cost of the COVID pandemic, but a systemic failure at the top of government caused by an inability or refusal to think strategically, plan structurally, and act collectively. For many years strategic foresight has been abandoned for political expediency which is now catching up with a Europe wholly ill-prepared for the crises with which it must contend, with Britain to the fore. In such a global context Brexit is irrelevant especially if London is unable or unwilling to act upon it beyond the rhetorical.

The future defence of Europe will rest ultimately not on the Americans, but Europeans, most importantly the three most powerful Europeans—Britain, France, and Germany. That is the one truth that is self-evident, whatever the Americans decide or whatever is imposed upon them. That truth includes a potentially devastating weakness. The closer one gets to the Russian border the stronger the commitment to deterrence, defence, and resilience; the further one moves away from the Russian border the weaker that commitment. Whilst Britain, France, and Germany remain leading economic and defence powers, representing almost 70% of all European defence investment and over 80% of defence research and development, all three have retreated from strategic realism into strategic pretence, with all of them subject to profound constraints over the use of power, that are as much cultural and political as actual.

It is this gap between strategy and realism, the almost wilful rejection of hard power, that Putin perceived prior to his invasion of Ukraine. It is the habit of making grand pronouncements that are never fulfilled allied to making defence investment pledges that are rarely met. In the dog-eat-dog world that the likes of Putin and Xi are forging, such pretence is lethal for those caught in between. Ruling elites in Berlin, London, and Paris seem too often to prefer playing at power, but are unable to make the tough choices power demands of them. This addition to the appearance of virtue has become the leitmotif of a false "strength," preventing the considered practice of power, and thus helping to destroy the very rules-based order they claim to champion.

Britain's political class long ago abandoned sound strategic judgement in favour of short-term self-interested politics, whilst much of London is locked in a false narrative of decline that also appears to implicitly accept the future demise of the United Kingdom. The Germans are still lost in a strategic desert of their own making even as the *Zollverein* (customs union) they created around themselves begins to dissolve. *Zeitenwende*? *Zeitenwas*? Power brings with it responsibility to self and to others, but for all the Euro-rhetoric Germany expounds Berlin lacks the essential solidarity with other Europeans vital if the European Project is ever to be more than empty words. The Americans, long the pillar of European security, are mired in a growing crisis about their future role in the world that is casting a harsh light on Europe's power pretence. Sadly, the Americans are too internally distracted and externally stretched to offset the incompetent retreat from power, will and strategy of the major Europeans. In any case, why should they?

The European nation-state remains the epicentre of European power, even if many Western European nation-states are now "led" by people "made weak by time and fate." Some seem no longer to believe in the very states they lead, just when Putin and Xi are reasserting hard power by hard states as the hard currency of power in the twenty-first century. It is *this* grand-strategic European challenge that Britain must first address, not assertions of a Global Britain as empty as the Treasury's coffers.

Integrated Review 2021, Integrated Review Refresh 2023 and Defence Command Paper 2023 have much to be commended, with innovative thinking built into their DNA, but they simply did not go far enough and were, à la Rommel, seemingly never meant to be read. Critically, the vision of a US-friendly interoperable high-end British future force whilst implicit in all the strategies is never explicitly laid out. IR 2021 also considered security in the round, i.e., the effects Britain needs to generate from the entirety of its security investment and the role of defence therein. The key word was "integrated," the use of all national means including defence to secure Britain and its interests.

The reason ultimately why the Americans are the ever-bigger elephant in London's room is the lack of sufficient investment in the Royal Navy, the Royal Air Force, and above all the British Army and the roles and missions London expects them to undertake. There is now a very real danger the British will continue to do what they have been doing since Suez, hedging between America, France, and Germany and assuming the Americans will always be there and always willing to afford London its strategic hubris. A long time ago the Americans were "over here, over paid, and over-sexed." Today, the Americans are still here, underpaid but over-taxed by Europeans who have become addicted to free riding.

The strength of London's thinking is that it correctly envisions security, intelligence, influence, and defence in the round and attempts to understand and respond to the changing character of the new hybrid, intelligence, and cyber "permawar" that Britain not only will fight but is fighting now. The problem is that it does not read let alone act upon its own analysis. It must. Russia might be down, but it is certainly not out, particularly as a hybrid and cyber warrior, and is rebuilding much of the force lost in Ukraine since 2022. China is not Russia and may be reasoned with, but Xi's Great Wall of Steel speech still reveals a Middle Kingdom only just beginning its Long March to geopolitical power.

Though Britain's public finances are important, they are not the first line of defence, but rather an enabler of defence. In the March 2024 Budget the government refused to further increase the defence budget, which revealed the extent to which London still sees defence as discretionary expenditure even in times of crisis. Rather, the Treasury stuck to the formula of £4.95 billion in 2024 and 2025 and £2 billion per year of additional funds out to 2030. The commitment to 2.5% by 2030 partially affects the problem but given defence cost inflation, the force will at best simply do little more than stand still, ill-equipped, under-manned and under-gunned. As former Secretary of State Gavin Williamson said,

> What is becoming increasingly clear is that the threats that we face need and require Britain and its allies to step up what it does in terms of building both capability and mass within our Armed Forces. That is going to require additional money to grow the size of our Army, Navy, and Air Force. Without doing that, we will be ill-equipped to face the challenges that our enemies are increasingly presenting us with.[2]

Sometimes in a country's history defence must simply be afforded, however expensive and however inconvenient. Soft power is cheap power and cheap power is no power. Hard power is and always will be the essential commodity in national influence, the funding of which must not be regarded as what is left over after all the other instruments of power—economic, diplomatic and development—have been afforded, or the excessive demands of welfare met. Even though defence is the fifth largest expense on the British public purse, it is simply not enough to meet the clash of wills with autocratic powers such as China and Russia, let alone all the other threats confronting the British, from the inevitable return of terrorism through to the dark side of globalisation and mass migration and water wars. In December 2023, the National Audit Office identified £16.9 billion over the next ten years of unfunded defence commitments. Unfunded? Surely a better use of language would be "chosen not to fund." Even were that money to be found stuffed behind an old Treasury Chesterfield it would only return defence investment to its baseline. To not find the money is to appease a dangerous reality and that is an old British story that never ends well.

A Return to Strategy

This book, *The Retreat from Strategy*, has considered the place of Britain's armed forces in British security policy. More than that, it has endeavoured to place them in the wider context of British grand and national strategy at another inflection point in Britain's long story as an independent major power. The essential message of the book is that Britain is again first and foremost a European power after a long detour as a global power. Britain is also one of three major European powers that cannot hide from the responsibilities that power imposes. Since 2001 Britain has either been fighting dangerous enemies far away or dealing with political, economic, and even a major health crisis closer to home. That heady mixture of crisis and failure has seen the London establishment retreat into a culture of declinism bordering on despair exacerbated by profound contentions over just who the British are and its place in the world. In fact, Britain is no longer declining from the peak of its one-time ascendency, it declined from there a long time ago. Equally, adjusting to life after exceptionalism is never easy for any state—just look at Russia.

As such, this book is about Britain today and tomorrow and the role it should aspire to play as a liberal democracy with a population of some seventy million souls, an advanced economy worth over $3 trillion, situated off the northwest coast of Europe at the beginning of an age of strategic and geopolitical tension. The book is called *The Retreat from Strategy* precisely because, given so many crises, government has become a fire-fighter engaged in a form of reactive whack-a-mole that has killed the generation and application of still immense power in pursuit of high national interests. That retreat has been accelerated by the profound confusion between values and interests at the very head and heart of the British government machine—London. This confusion has led the British into several campaigns in places such as Afghanistan, Iraq, Libya, and Ukraine which have failed precisely because values and interests were not aligned. Even if Britain's values are deeply offended, if there is not also a threat to critical British interests then sooner rather later political and popular support falters and Britain's commitment fails.

There are thus several takeaways from this book. First, Britain should only ever seek to act as a force for good if its values and interests are properly aligned and if it is willing and able to pay the price of any action. Second, Britain's immense soft power is all very well and good in a peaceful age, but at times such as these the residual strategic brand which the British still possess will see much of the world expect London to act. To paraphrase Hobbes, political promises about strategic goals with insufficient means are but words and of no use to any man or woman. Britain needs not just armed forces but hard fighting power commensurate with its development, economy and technology which is simply not the case today. Third, politicians must stop treating citizens like children and routinely exaggerating their own ability to shape events and Britain's capacity to protect itself. Britain's future security and defence will rest on a new balance between power projection and people protection. Such a balance will only ever be realised in partnership between power and people. Fourth, politicians must stop seeing sound defence as a cost rather than a value. And it is here that a return to strategy will begin. Britain cannot afford to do all things all the time in pursuit of its own security, let alone that of allies and partners. Like all medium-ranked powers, strategy is more not less important to Britain than it is to the mighty US, although clearly it would help if the Americans sorted out their own confusion between values and interests. This is ultimately because strategy is about choices—hard choices. It is the method by which priorities are established, the framework within which the critical and essential interests are distinguished from the hoped for.

Strategy itself must be partitioned given the immense power and many levers a powerful state like Britain must exercise to be fit as a power. Britain may not have a stated grand strategy, but government must be able to think grand strategically and coherently. Only then will what really matters to Britain in this world be understood and national strategy—the application of said British means—be properly applied. Since 2010, precisely because London has been so unsure of itself and Britain's place, all the British high-strategic documents have too often read like one of those glossy magazines one gets when flying that say nothing and no one ever reads. If they do, it is only to look at the last few pages where there might, just might, be some useful information.

At a practical level, thinking grand-strategically will require trained statesmen and stateswomen, with very senior cabinet ministers educated in their respective missions and tested to that effect. Senior appointments in government also need to include many more influential and better trained strategists and defence specialists, particularly in the National Security Council staff, as well as a national command and coordinating apparatus that vastly improves on COBRA. There is also a cultural component. Too much emphasis is still placed in high levels of government on being "one of us" and coming from the "right" background, whilst lower down the command chain too much emphasis is placed on diversity above competence. Indeed, only when British high strategy is clear to those responsible for it, and they are equipped to lead Britain and believe in Britain, will the vital relationship between ends, ways and means be re-established.

Take the army. Size matters, with credible and relevant fighting power and readiness the real measures of strength. Ukraine has 100 manoeuvre brigades and is only just holding a defensive line, whilst Britain can only deploy at best several such brigades. At the very least, Britain must have an expeditionary army that can enter the continent in some strength together with sufficient and robust logistics. In the face of a Russia committed to aggression and given NATO's increased defensive frontage, the British Army should be able to field at least one full corps headquarters and two heavy divisions, one at thirty days readiness and the other at ninety days, in addition to Special Forces and other highly mobile strike forces together with their deployed defence and enablers. That is not the case today. The armed forces must also be big enough and capable to afford Britain a seat at the table in Washington and the institutions vital to London's defence strategy.

Unfortunately, London has no clear focus on how to defend the realm, leaving British defence policy incoherent and purposeless, with little relation between defence strategic ends, ways and means. There are other consequent critical lacunae, such as the lack of reserves, all of which since 2010 have been committed to cover personnel shortfalls, and no industrial and technological base to support the armed forces in general. Worse, there is no real link between the size and capability of the armed forces and the need to defend an island that cannot feed itself should trade be disrupted or play an appropriate role in NATO land deterrence and defence.

Put simply, at current levels of defence expenditure Britain can no longer afford both a bespoke continually-at-sea nuclear deterrent and credibly and appropriately powerful conventional armed forces that meet its NATO commitments. This is partly due to defence cost inflation, but also due to poor returns on defence investment caused by gold-plated procurement. Britain also has wider defence-strategic commitments, such as the AUKUS pact, which whilst offering benefits such as enhanced influence with critical partners, also cost money reducing further the investment the armed forces need to realise minimum force goals.

What to do? All future defence reviews, whatever they are called, must depart from recent such exercises, and begin with an honest analysis of likely global and European security challenges over the next ten years, central to which would be identification of the fighting power needed to confront Russia. Only then can Britain deduce the minimum capabilities needed in both the nuclear and conventional domains to dissuade them, with or without American support. Finally, as mentioned earlier, the Chief of the Defence Staff should become the Commander, UK Armed Forces (CUKAF) and unequivocally be their strategic commander with all the authority this implies. All electronic, cyber, space and information warfare should be merged within a revamped Strategic Command which should be re-titled Joint Command.

There is a vital need for wholesale reform designed to drive jointness deep across all the armed forces and their civilian masters. It must start at the top with the wholesale reform of the National Security Council but above all the Ministry of Defence. Above all, London must stop lying to itself about the state of the British Army because only then will NATO allies believe that Britain can deliver the fighting power which is central to the NATO Defence Planning Process and land-centric deterrence.

Until London's body politic is once again aligned with Britain's minimum grand strategic interests the ends, ways and means of British military power will continue to be out of sync. For too long the Royal Navy, British Army and RAF have been seen as a cost that can only be afforded if Britain is prosperous enough, rather than a critical value that must be afforded. These tensions have been compounded by senior military commanders who bet that technology can offset mass whilst also trying to meet an expanded task-list with an ever-smaller force. This has placed pressures on experienced personnel who are leaving the army in significant numbers. Many allies value the quality of British officers but if the army continues to erode that quality will erode with it.

Why does all the above matter? It is because at some point a relatively small number of British citizens—Tommies (both male and female)—will find themselves on the wrong, very sharp and potentially lethal end of bad British strategy. It is a reasoned assessment of this book that Britain does not spend enough on defence and what it spends it does not spend wisely. Worse, too many in office and in power (the two are not the same these days) spend too much time deceiving themselves and others that somehow ends, ways and means are balanced out.

For Britain, the only way to both provide credible security and defence and afford it is with a laser-like and unequivocal focus on meeting its NATO obligations, whilst in the process freeing up US capability to focus on China on the West's collective behalf. Hard choices? Take nukes. If London really does want Britain to retain an independent nuclear deterrent *and* appropriately powerful and credible conventional forces sufficiently capable to help keep the threshold for the use of nuclear weapons in Europe high, then it is going to have to spend a lot more money on defence as well as restructure the Ministry of Defence and give the service chiefs the authority they need to make military-strategic decisions. The alternative? Belgium with nukes!

Heroic Hearts?

British strategy must have four interest-led essential elements.

First, keep the US engaged in European security and defence by demonstrating to Washington that Britain is prepared to lead Europeans in serious defence investment. This will also help ease American over-stretch by keeping the US strong where it needs to be strong through an equitable sharing of burdens.

Second, establish a Euro-strategic partnership with Germany that recognises Germany's strategic role in European economic and political stability, in return for Berlin recognising British and French leadership of Europe's military effort. France will resent German leadership and further resent any relationship between London and Berlin that might appear to eclipse the Franco-German axis. London will need to work hard to overcome French suspicions.

Third, maintain the Franco-British strategic defence partnership with a particular emphasis on a joint effort by London and Paris to improve and increase European expeditionary military capabilities, as well as achieving greater synthesis between their respective nuclear weapons programmes.

Fourth, re-establish and restore relationships with the English-speaking Commonwealth (Australia, Canada, et. al.) to reinforce a Congress of Democracies as central to a new West, a world-wide web of secure democracies.

Sir Tony Blair is clear about the vital need for considered strategy:

> To do this [strategy], in the highly pressured world of contemporary politics, requires taking real time to stand back and get the clarity which is key to any successful strategy. It requires debate, Socratic-style exchange of opinion, testing and refining. It can't be snatched at towards the end of a busy day. It has to be the day's work.[3]

Britain must also urgently re-consider in the round (and properly) policy, strategy, structure, forces, and investment if London is to be credible as a shaper, deterrer and actor in the face of coming challenges.

If London continues to strip, denude, and cut the instruments of national power simply to placate the politics of the short-term, the costs in the long-term will grow exponentially. Indeed, only by properly investing in effective diplomacy and a powerful military will Britain be able to influence the very levers London uses to exert what grand strategy it can reasonably exert: Washington, the EU, NATO, and the UN. Britain long ago abandoned its position at the centre of affairs and will thus need to make a concerted strategic effort to re-establish sufficient influence over these centres of power so that once again Britain can multiply influence via leverage. Britain faces a choice: invest in the power necessary to achieve both the means and ways of strategy, or face the consequences of shattered ends.

London must move decisively to correct the narrative of self-decline because it is undermining any chance Britain may have to prepare properly for the defence of its legitimate future. To that end, Britain is in urgent need of a strategic narrative and only a sound British grand strategy can provide such a narrative. All states use history to create a sense of patriotism. Indeed, it is the identity of the people with the state that is the quintessence of the state. Unfortunately, the assault by the political left on British identity, and the profound ignorance of history by much of the political right, has further undermined efforts to reinforce Britain's historical narrative with a strategic narrative. The result is that the story of Britain's role in the world is polluted daily by attempts to impose twenty-first-century values on eighteenth- and nineteenth-century actions, often to shame the British into inaction. The result is the artificial acceleration of national decline caused by guilt by distant association built on a corrupted view of history.

Whilst misadventure and criminality can be found in British history, the British were amongst the most radical of liberal reformers and remain so today. Indeed, it was the merger of British power with Victorian liberalism that led to the creation of much that is good in the world of today. History is always comparative, and comparatively the British contribution to peace, stability and justice has been outstanding. A British strategic narrative must reflect that for without it a consensus over British national strategy will be very hard to forge in an increasingly diverse and uncertain society. Decisive political leadership is needed to that end.

Sadly, the very process of strategy-building has become part of Britain's problem. Strategy in Britain is too often an exercise in bureaucracy, and ultimately, futility. Indeed, if the National Security Council is to realise its potential it is vital Britain's leaders take British strategy back from the bureaucrats. It is also vital Britain's historical story is taken back from the historical revisionists and doom-mongers who corrupt history for their own nefarious and anti-British ends. In other words, it is not just threats and challenges that must be confronted, but the very culture that the modern British elite have created. Fail to do so, and the national cohesion upon which strategy and policy is reliant will be dangerously partial and incomplete. Strategy-lite policies lead too often to a tendency to treat citizens as children, comforting them rather than securing them, part of a rapid descent into strategic denial that makes it hard often to discern linkages between political intent and government action.

Above all, the British need to be very clear about the whys and wherefores of their national strategy because selling it to a sceptical public that has lost faith in the political class will prove challenging. The message should be clear: to be legitimately secure Britain must play a full role in helping to guide the world safely through a dangerous period of profound strategic transition and help shape that transition in the British interest. To meet such a challenge Britain needs the required hard power it can generate as part of a collective effort by the democracies. No ifs, no buts! Too often the implicit message from a London paralysed by political correctness is that British grand strategy is simply to keep the international system warm until the likes of Germany, China, and India take over—and that since Britain is no longer that which it was it cannot be of any consequence. This is complete and utter defeatist nonsense. Britain must compete.

Successful transition from a British viewpoint will be achieved through the embedding of new power through old and new partnerships in functioning institutions central to the system of global security governance that the West created. It will be achieved by London actively reinforcing the legitimate state as the focal point for twenty-first-century security and identity, even as the old post-colonial order is being torn down across large swathes of the world by radical anti-state elements. That means a London committed to a credible British ability and capability to engage in all forms of co-option and coercion because, like it or not, Britain is engaged in an existential battle.

The whole question of where best to invest in security will also need to be revisited if the home base is to be protected given the necessarily complex layers of defence that contemporary security demands. Such security must combine effective intelligence, kinetic layers of defence, protection of critical infrastructures and cyber-defence. Security will also require a robust and resilient society sufficiently coherent to understand the risk and willing as a national community not only to pay for its security and defence, but to play an active role in it. Indeed, a secure home base is the *sine qua non* of power projection. And Britain will need to project power.

Therefore, the successful crafting and execution of effective *national* strategy in the contemporary age will likely require more than \ penchant for incrementalism. There are significant parts of Britain's security and defence architecture that will need to be rethought if Britain is to prepare effectively to confront a radical age. "Strategic" will need to mean strategic.

James Harrington once wrote,

> No man can be a politician, except he be first a Historian or a Traveller, for except he can see what must be, or what may be, he is no politician ... but he that neither knows what has been, nor what is, can never tell what must be, nor may be.[4]

This is a big strategic era and Britain's politicians could all too predictably blow it if they are too small for the challenge. Strategic shrinkage, and with it Britain's reduced capacity to deal with challenges, is a fact and must be addressed. At the very least many of the assumptions concerning Britain's strategic interests must be considered and reconsidered in the light of the exceptional change that is taking place in Europe and the wider world.

Equally, strategic planning never takes place in an entirely free space, as Sir Hew Strachan has pointed out.[5] There are always a host of enduring commitments that must be upheld. As Sir Lawrence Freedman recognised, national strategy is also a perpetually moving target:

> Strategy is often expected to start with a description of a desired end-state, but in practice there is rarely an orderly movement to goals set in advance. Instead, the process evolves through a series of states, each one not quite what was anticipated or hoped for, requiring a reappraisal and modification of the original strategy, including ultimate objectives.[6]

In other words, strategy is dynamic and those who craft it must be agile enough and capable enough to shape such change, not have it imposed upon them.

Given all the above, the principal aim of British strategy is the same as that of any other leading state—safeguarding the nation, the economy and the people in this world and tomorrow's world, not yesterday's world. Sadly, the retreat from strategy that has affected and afflicted London is now seen in an essentially defeatist, declinist and by and large ignorant British political class, too often seeking solace in political clichés that reflect their own lack of understanding of how power and strategy really work. Despite the undoubted change that is taking place in the world, London also routinely exaggerates the capability and ability of the emerging powers to shape the strategic environment, and underestimates its own power as an alibi for unforgiveable unpreparedness and inaction.

There is an alternative British reality and future. Whilst most of the British people still believe in Britain as an independent power in the world, many of those at the head of Westminster politics and London bureaucracy simply do not. Britain's declinist elite are simply latter-day heirs to the French elite as Churchill described them in the aftermath of the First World War:

> Worn down, doubly decimated, but undisputed masters of the hour, the ... nation peered into the future in thankful wonder and haunting

> dread. Where then was that SECURITY without which all that had been gained seemed valueless, and life itself, even amid the rejoicings of victory, was almost unendurable? The mortal need was Security.[7]

London must finally begin to stand up for Britain and at least suggest they believe in Britain and its people.

There can be no question that with the right political will Britain could do far more, even alone and even with minimal international support, were domestic opinion confident that its leaders were up to the task and prepared to accept the costs and sacrifices of action. At the very least, the British could do more to place themselves again at the centre of the West. The alternative is security pretence, isolationism and decline, into which much of Europe has already fallen. Self-delusion may be comforting in the short-term, but it is dangerously misguided over the medium to long term.

Why Britain Matters

Aldous Huxley once wrote that, "human beings have an almost infinite capacity for taking things for granted." *That* is the real danger of Britain's retreat from reality. Britain must once again grip the big picture as it has often done in the past! Britain *is* at a vital strategic crossroads. The return to strategy is not simply about whether London can exploit its still potent strategic brand or still very consequent power. The return to strategy is ultimately about the fate of Britain itself: whether Britain goes into terminal decline and eventually breaks up, or finds its rightful place as a medium-sized but influential European power in the twenty-first-century world. If Britain's political class continues to talk the talk of power but walk the walk of irresolution and weakness that they have trodden for far too long, then Britain's future is not only uncertain but possibly disastrous.

So, why does it matter? Since the end of the Cold War Britain has reacted to crises that offend London's values, but do not really threaten its vital interests. Then, since British vital interests are not at stake, unless the response is short and relatively costless London soon loses interest for fear a commitment will become politically damaging. Contemporary history is also replete with examples of London hiding from the responsibilities of its own power under the delusion Britain can stay out of a war only then to be dragged in wholly unprepared and ill-equipped, as the British armed forces are today. Above all, the consequence of such strategic illiteracy is that the people who must close the ends, ways and means gaps with their lives are the young men and women of the British armed forces—Tommy!

As Kipling wrote all those years ago:

> You talk o'better food for us, an' schools, an' fires, an' all: We'll wait for extry rations if you treat us rational. Don't mess about the cook-room slops, but prove it to our face. The Widow's Uniform is not the soldier-man's disgrace. For it's Tommy this, an' Tommy that, an' "Chuck him out, the brute!" But it's "Saviour of 'is country" when the guns begin to shoot; An' it's Tommy this, an' Tommy that, an' anything you please; An' Tommy ain't a bloomin' fool—you bet that Tommy sees![8]

In his April 2024 statement then Prime Minister Rishi Sunak said, "Today is a turning point for European security and a landmark moment in the defence of the United Kingdom. It is generational investment in British security and British prosperity, which makes us safer at home and stronger abroad."[9] All well and good but Sunak also quoted Churchill who in 1934 said, "To urge the preparation of defence is not to assert the imminence of war. On the contrary, if war were imminent preparations for defence would be too late."

If Britain is to rebalance the ends, ways and means of its defence and its wider defence influence, London will need to return to strategy. To do that, Britain will need statesmen and women in 10 Downing Street, not managers or social workers, and similarly minded people in a revamped and powerful National Security Council armed with a new executive arm. Why? Because the world is simply too dangerous for politics to pretend to be strategy and critical interests to be confused with vague values. Some ages forgive mediocrity. The coming strategic age will be no such age. Machiavelli wrote in *The Prince*, "All courses of action are risky. So, prudence is not in avoiding danger (it is impossible) but calculating risk and acting decisively. Make mistakes of ambition, not mistakes of sloth. Develop the strength to do things, not the strength to suffer."[10] Which highlights the greatest need of all: for Britain's high political and bureaucratic elite to have the will and the imagination to think the unthinkable. For, as Plato said, only the dead have seen the end of war. There is plenty for the new British government to ponder.

Think about it, London!

SCENARIO

BRITAIN DEFENDED

In 2031, China and Russia launch a full-scale hybrid and cyber-attack on Britain. The attack fails because of the 2025 decision to create an integrated defence across intelligence, cyber and kinetic security, and defence allied to enhanced resilience of critical governance structures, infrastructures, and people. Britain absorbs the strike and hits back at vulnerable Russia in the first systemic cyberwar. NATO's adversaries are unable to sink Europe's unsinkable aircraft carrier, and NATO's forward defence holds. Russian plans to seize rump Ukraine and much of the rest of Eastern Europe are in disarray. The key to victory? The return to strategy in 2025 and London's coherent security and defence posture that balances efficiency and effectiveness across government and wider society.

War!

On 9 October 2031 two flotillas of frigates and destroyers from both the Russian and Chinese fleets break off from Exercise Tsentr 35 and move to Svalbard to "assist in the rescue and recovery of the *Admiral Amelko* which was damaged in an explosion and to establish a screen to protect the fishing vessels."

Crucially, they carry the latest Tsirkon hypersonic anti-ship missile. Two Russian Laika-class nuclear attack submarines (SSNs) also head towards the Norwegian coastline to prevent the Norwegian Navy from deploying into the region. A Russian carrier task group and amphibious forces also move south, detach near Svalbard, and prepare for an amphibious landing.

On 10 October, both Russia and China accuse Norway of destabilizing the region and acting aggressively, whilst Russia blames Norway for the "sinking" of the *Admiral Amelko*. It is fake news. There is widespread outrage in Russian media even though Washington reveals satellite photos of the *Amelko* at anchor at the Severomorsk fleet base. China also condemns Norwegian actions and cautions NATO following the swift activation of British and Dutch naval and amphibious forces.

On 11 October, Littoral Response Group North (LRG (N)) begins to move into position several hundred kilometres south of Svalbard, as it prepares to insert teams onto the islands to link-up with the Norwegian military. The North Atlantic Council, meeting in emergency session, demands that Russia and China step down their forces. The NATO hope is that the arrival of the UK-led amphibious force will be enough to deter any further aggressive acts by China and Russia and de-escalate tensions. The UN and EU also seek to calm the situation and separately propose an international-led investigation into the explosion on the *Admiral Amelko*. It is also hoped that by establishing a presence off the coast of Svalbard it will enable a stronger NATO force to be deployed to the region.

However, as the crisis intensifies Chinese and Russian cyber and electronic warfare units also begin to target key infrastructure and communications capabilities in Svalbard and on mainland Norway. They also try and blind US, British and French surveillance satellites, in the hope of drastically reducing the notice of impending Russian and Chinese actions. However, crucially all these systems have been hardened against just such an attack. At the North Atlantic Council Germany questions the deployment of a major NATO combat force stating that Berlin's lawyers do not believe an attack on Svalbard would trigger an Article 5 contingency, but still says that its forces will participate in NATO operations.

On 12 October, the aircraft carrier HMS *Queen Elizabeth*, at the core of the UK's Carrier Strike Group, departs Portsmouth ahead of schedule and moves northward at speed towards the Arctic region. The Dutch amphibious HNLMS (His Netherlands Majesty's Ship) *Rotterdam*, with a company of marines aboard and a contingent of escort ships, also departs Den Helder and is directed to link up with the *Queen Elizabeth* Carrier Strike Group. The US also increases aerial surveillance in the Arctic, whilst a US Marine Expeditionary Unit currently operating in the Mediterranean Sea is redirected north toward the Norwegian Sea.

Satellite coverage over the Arctic region together with air-breathing systems provide an effective intelligence, surveillance and reconnaissance picture. This enables the Alliance to anticipate China's sudden unveiling of a hitherto undetected and complex web of anti-access/area-denial (A2/AD) systems along the Greenland coast and the southern tip of Svalbard. Chinese forces on Svalbard also reveal themselves and begin to deploy mobile advanced anti-ship ballistic and cruise missile systems on the south of the island. However, US special forces are already in place and sabotage many of the facilities. There are still enough Russian and Chinese "smart sensors" allied to a network of "satellites" and aerial surveillance for them to attempt what the Americans call "the intelligence grab." Russia also declares an air-defence identification zone over the whole of the Arctic and begins to challenge the aircraft of all other countries operating in the region.

13 October 2031

And then it starts. Early on 13 October key early-warning systems on Svalbard are attacked with direct fire and munitions delivered by a swarm of drones. The Norwegian Navy responds by deploying the frigates *Otto Sverdrup* and *Thor Heyerdahl* to southern Svalbard. Russian intelligence reports the Norwegian 1st Submarine Squadron is also preparing to deploy the super-quiet *Ula*, *Utsira* and *Utstein* which could pose a significant threat to the amphibious force. To coerce Oslo, Russia responds by activating several brigades, tanks, artillery, and surface-to-air missile capability to the West of Murmansk, close to the Storskog border crossing but stop just short of entering Norwegian territory. For the moment they then dig in.

0001 hours: A Russian nuclear attack submarine detects the Norwegian flotilla moving towards Svalbard and is ordered by the Russian Arctic Command to strike and sink the frigate *Otto Sverdrup*. What the Russian commander does not know is that close to the Norwegian frigate, *HMS Agincourt*, an Astute 2 nuclear attack submarine, is lurking.

0100 hours: Multiple cruise missiles strike facilities in Bardufoss, where US Marines, Norwegian military, and UK Royal Marine Commandos have been conducting tri-lateral training. Forewarned, they are not there and are already en route to Kirkenes to block Russian forces near the Storskog border-crossing, who are preparing to sweep through northern Norway along the Kirkenes-Narvik axis.

0110 hours: Chinese and Russian military action against Svalbard begins together with sustained air-strikes.

0300 hours: Russian and Chinese naval infantry advance forces land but are surprised to find a significant force of US, Royal

Navy and Norwegian Marines already in situ who force them back. Russian and Chinese follow-on forces are also thwarted with several of the ships sunk. Chinese and Russian Special Forces also fail to secure the southern peninsula and lose control over the sites where they had intended to install advanced A2/AD systems. Meanwhile, hundreds of Russian paratroopers appear over Svalbard's coastline charged with quickly securing ground and establishing beachheads so that additional A2/AD and other missile systems can be disembarked. They are destroyed by F-35 fighters operating from HMS *Queen Elizabeth*.

0300 hours: Attacks also take place on Britain's undersea telecommunications network and gas and oil pipelines from Norway. Even though London immediately declares a major emergency, redundant and hardened systems prevent the collapse of reinforced critical infrastructure, including hospitals and police, which are protected by an elaborate civil cyber defence.

0315 hours: As soon as the force commander receives notification that the Russian military has invaded, LRG (N) orders cruise missile attacks on several S-400 missile systems and additional military capability being offloaded on the south of Svalbard. Russian Special Forces (Spetsnaz) are again beaten back by US Delta Force and Britain's Special Air Service as they try and seize key infrastructure. Suddenly, the extensive A2/AD "system of systems" Beijing and Moscow had banked on being in place along the southern Svalbard coastline and linked to similar systems established on the east coast of Greenland and on Franz Josef Land is destroyed. They begin to lose control of events.

0325 hours: LRG (N)—protected against anti-ship cruise missile attacks by a mix of AI-enabled, autonomous and cyber systems which begin to paralyse local enemy command and control—moves closer to its objective, enabling it to deploy more strike teams.

0330 hours: Article 5 is triggered by the North Atlantic Council and further US, Canadian, Swedish and Finnish special forces move to support British, Dutch, and Norwegian forces in the region and the fight on Svalbard. They find and strike Chinese and Russian high-value targets on Svalbard and open a corridor for a rapidly assembling heavy US force.

0450 hours: Royal Marine Commandos, on "strip alert" aboard HMS *Albion*, are then deployed to Svalbard, with several strike teams conducting a hazardous 50 km ship-to-shore movement to land on the southwest and northwest of Svalbard.

0530 hours: Covert insertion of the first wave of multiple teams is successful, and is followed by a second wave by air as Chinese and Russian forces lose the capacity to track and target them.

0700 hours: Multiple strike teams conduct infiltration from the coast of Svalbard into the interior. Some move to link up with Norwegian forces, whilst others begin the mission to find and destroy high-value targets, primarily more A2/AD systems. This first action is considered essential to allow follow-on forces into the region. They achieve rapid success, supress Chinese and Russian A2/AD, and allow the arrival of US forces to serve as a catalyst for so-called off-ramp negotiations.

0710 hours: HMS *Albion* departs the area after deploying strike teams.

0845 hours: Following their covert insertion, the strike teams have a disproportionate effect, quickly finding and striking high-value targets on the island using their strategic capabilities and drone-delivered munitions from distance. Russian and Chinese Special Forces on Svalbard are also located and destroyed.

0900 hours: Chinese and Russian forces try and launch a major maritime-amphibious operation on Svalbard supported by air power and protected by the remaining A2/AD systems. However, they are intercepted by American and British submarines and air power and destroyed.

0915 hours: Russia also tries to launch a series of "hunter-killer" operations using helicopters in conjunction with ground forces to find and destroy UK strike teams operating on the islands, but they too are detected early and destroyed. Thereafter, Russian find-and-destroy operations prove to be highly ineffective. The operation continues with further "hunter-killer" aircraft deployed to locate and strike NATO forces on Svalbard. Attacks on A2/AD systems are halted.

1400 hours: The HMS *Queen Elizabeth* carrier strike group also erodes the wide weapons engagement zones Russian and Chinese forces have sought to establish, with close air support of ground forces forward deployed.

1500 hours: As the momentum of NATO attacks gathers, LRG (N) merges with the HMS *Queen Elizabeth* Carrier-Strike Group to form a hybrid Expeditionary Strike Force (minus LRG (S) which is still on its way back from Suez). This enables LRG (N) to offer reinforced logistical support to strike teams on the island and to redeploy forces in response to the last elements of the enemy A2/AD threat.

1700 hours: US-led air power operating from bases in the UK, supported by the F35B Lightning 2 strike aircraft on HMS *Queen Elizabeth*, strike Chinese and Russian A2/AD capabilities on Svalbard. Over seventy Chinese and Russian personnel are killed, and several mobile A2/AD systems destroyed.

1900 hours: Despite the "shock-and-awe" tactics of Chinese and Russian forces their attack has failed with both north and south Svalbard islands firmly under NATO control. S-400 anti-aircraft missile systems with an operational range of 400km have either been blocked or destroyed throughout the archipelago. The layered and integrated web of Chinese and Russian A2/AD systems that threatened to stretch from Greenland through Svalbard to Franz Josef Land, blocking all air and sea approaches to the Arctic and threatening to impose a disproportionate cost on NATO follow-on forces, is no more. The Chinese and Russian carrier groups held in reserve in the South Barents Sea designed to provide close air protection to cover the offload on Svalbard, as well as protecting vital ground forces, now begin to withdraw as the threat of NATO counter-fires increases. D-Day in the Arctic is effectively over.

Britain and Norway are secure, whilst Finland and Sweden with their Total Defence Concepts applied to the full have beaten off Russian cyber and information warfare. Crucially, the information warfare the Chinese and Russians had launched against Britain in parallel with their cyber-attacks also failed because Moscow and Beijing had failed to realise how adaptive Britain's critical infrastructure was.

The real object of the Svalbard attack is revealed: the Baltic States, as well as control of the North Cape and the Arctic. The campaign had been designed to distract the Alliance and force it to move significant land forces to the north to cover Finland, Norway, and Sweden. However, the build-up of Allied land forces between 2025 and 2035 enabled SACEUR to deploy a sizeable blocking force opposite Russia's Western Military Oblast supported by massive air and missile power. Put simply, it was NATO that succeeded with the "intelligence grab" that had been key to Russian and Chinese planning.

15 October: The Allied follow-on force arrives and prepares for strikes and, if ordered, forcible entry onto Svalbard. However, with Chinese and Russian forces rapidly degrading, Allied commanders quickly realise that they can secure their objectives with

the forces already deployed. The Chinese and Russian anti-access threat has failed as commando strike teams are successful in locating and targeting the final high-value targets. This allows the US Navy and Royal Navy to come close to Svalbard in sufficient strength to allow effective targeting of the integrated air defence system by the US Air Force.

Having been blocked on Svalbard, Russia and China now seek to de-escalate tensions in the Arctic before the situation spirals out of control. The Sino-Russian strategic partnership has failed to secure critical sea lines of communication and control over the vast resources of the Arctic. It has also failed to gain control over the access to the Bering Sea, thus greatly inhibiting the ability of Russian nuclear submarines to operate far into the North Atlantic.

China had hoped the "war in the north" would stretch US forces so thin that Chinese forces could finally move against Taiwan. But, deep in the South China Sea, the USS *Ronald Reagan* and its task group...

BRITAIN'S RETREAT FROM STRATEGY

ELITE QUESTIONNAIRE

David Richards and Julian Lindley-French are co-writing a book which seeks to answer two fundamental questions. First, what role should a country of Britain's size, wealth, traditions, culture, experience, allegiances, memberships, location, capabilities, and capacities seek to play in the twenty-first-century world? Second, is Britain playing such a role and if not, what grand strategy does Britain need to adopt to realise such ambition?

1. The book assumes Britain has retreated from strategy—both grand and national strategy. Is this correct, if so how?
2. Is there a need for a grand British strategy or simply the capacity to think grand-strategically?
3. Is "managing decline" a reality or simply a euphemism for a high political and bureaucratic establishment that is incapable of creating and acting upon sound grand and national strategy?
4. Strategy assumes the considered application of force and resource over time. Why is Britain so locked into the short-term at the expense of the longer-term?
5. Effective strategy assumes unity of purpose and effort at the high end of the government machine. Why is whole-of-government action so hard for British governments to achieve?

6. Successive British governments have announced strategies of all kinds for a range of vital issues, yet so little seems then to happen. Why?
7. Britain spends ever more on defence, but British forces seem to become ever smaller and despite claims to the contrary less capable. Why?
8. The Cameron government created the tools for sound strategy, most notably the National Security Council, but these tools have seemingly failed to deliver coherent [grand] strategy and effective implementation. Why?
9. Do the National Security Strategies and Integrated Reviews work? If not, why?
10. Should politicians and senior civil servants/diplomats not be trained in the skills required to devise and execute coherent strategy(ies)?
11. What armed forces should Britain aspire to have? What role? How big?

NOTES

PREFACE: THE SLEEPWALKERS

1. Eliot Wilson, “Why did it take Rishi Sunak so long to up defence spending?” *Spectator*, 23 April 2024.
2. Hugo Gue, “UK must ramp up military spending to 3% GDP, Shapps warns,” iNews, 24 December 2023, available at: https://inews.co.uk/news/politics/uk-ramp-up-military-spending-gdp-shapps-2823808?ico=in-line_link (last accessed 15/05/24).
3. *The Oxford English Dictionary*, Oxford: Oxford University Press, 1989, p. 1052.
4. “Defending Britain from a more dangerous world,” speech by the Right Honourable Grant Shapps, Secretary of State for Defence, Lancaster House, London, 15 January 2024, available at: https://www.gov.uk/government/speeches/defending-britain-from-a-more-dangerous-world (last accessed 16/05/24).
5. On condition of anonymity.
6. Laura Hughes, “UK locked down 3 weeks too late, Matt Hancock tells Covid inquiry,” *Financial Times*, 23 November 2023, available at: https://www.ft.com/content/f46183c5-d9c3-4b36-b8a6-856529d9c381 (last accessed 16/05/24).
7. In response to the book questionnaire.
8. In response to the book questionnaire.
9. In response to the book’s questionnaire and also at a lecture given to the Royal College of Defence Studies, March 2022.

10. Jerry Purnelle, online at: https://www.jerrypournelle.com/reports/jerryp/iron.html (last accessed 16/05/24).

INTRODUCTION

1. See Koda Yodi, "China's Blue Water Navy Strategy and its Implications," CNAS, 20 March 2017, available at: https://www.cnas.org/publications/reports/chinas-blue-water-navy-strategy-and-its-implications (last accessed 16/05/24).
2. See David French, *Army, Empire and Cold War: The British Army and Military Policy 1945–1971*, Oxford: Oxford University Press, 2012.
3. "Tilting horizons: the Integrated Review and the Indo-Pacific," House of Commons Foreign Affairs Committee, 30 August 2023, p. 4, available at: https://publications.parliament.uk/pa/cm5803/cmselect/cmfaff/172/summary.html (last accessed 16/05/24).
4. Ian Williams, "Tory floundering over China is a gift to Labour," *Spectator*, 19 July 2023.
5. In response to the book questionnaire.
6. In response to the book questionnaire.
7. "Global Soft Power Index 2022," Brand Finance, available at: https://brandfinance.com/press-releases/global-soft-power-index-2022-usa-bounces-back-better-to-top-of-nation-brand-ranking (last accessed 16/05/24).
8. "Top Countries by GDP," available at: https://www.worldometers.info/gdp/#top20 (last accessed 16/05/24).
9. Global Firepower, "2024 Military Strength Ranking" available at: https://www.globalfirepower.com/countries-listing.php (last accessed 16/05/24).
10. "Countries in the world by population," available at: https://www.worldometers.info/world-population/population-by-country/ (last accessed 16/05/24).
11. "Countries by Area," World Atlas, available at: https://www.worldatlas.com/features/countries-by-area.html (last accessed 16/05/24).
12. "UK trade in numbers," GOV.UK, 19 April 2024, available at: https://www.gov.uk/government/statistics/uk-trade-in-numbers/uk-trade-in-numbers-web-version (last accessed 16/05/24).
13. Raul Amoros, "Visualizing the State of Global Debt, by Country,"

Visual Capitalist, 1 February 2022, available at: https://www.visualcapitalist.com/global-debt-to-gdp-ratio/ (last accessed 16/05/24).
14. J. K. Galbraith, *The Anatomy of Power*, Boston: Houghton Mifflin, 1983.
15. Szu Ping Chan, "Net Zero will hamper the West and boost China, warns CEBR," *Telegraph*, 26 December 2023, available at: https://www.telegraph.co.uk/business/2023/12/26/net-zero-hamper-west-boost-china-warns-cebr/ (last accessed 16/05/24).
16. In response to the book questionnaire.

1. BRITAIN, STRATEGY AND HISTORY

1. In response to the book questionnaire and as part of a lecture to the Royal College of Defence Studies, March 2022.
2. Samuel J. Hurwitz, "The Roots of British Foreign Policy," *Current Affairs*, Vol 26, No. 273, 1964, p. 296.
3. See Helen McCarthy, *The British People and the League of Nations*, Manchester: Manchester University Press, 2011.
4. The Nassau Agreement was concluded in December 1964 between the US and Britain. Its ostensible purpose was to resolve the so-called Skybolt Crisis and the American decision to cancel a missile system for which Britain's airborne deterrent had been designed. After much negotiation the Americans agreed to supply the British with the then-state-of-the-art Polaris submarine-launched nuclear missile system.
5. For a full understanding of what John Reid meant see Anthony King, "Understanding the Helmand Campaign: British Military Operations in Afghanistan," *International Affairs*, Vol. 86, No. 2, 2010.
6. "U.S. Debt to GDP Ratio 1960–2024," Macrotrends.net, available at: https://www.macrotrends.net/countries/USA/united-states/debt-to-gdp-ratio (last accessed 16/05/24).
7. "U.K. Debt to GDP Ratio 1960–2024," available at: https://www.macrotrends.net/global-metrics/countries/GBR/united-kingdom/debt-to-gdp-ratio (last accessed 16/05/24).
8. Deborah Haynes, "US general warns British Army no longer top-level fighting force, defence sources reveal," Sky News, 30 January 2023, available at: https://news.sky.com/story/us-general-warns-british-army-no-

longer-top-level-fighting-force-defence-sources-reveal-12798365 (last accessed 16/05/24).

9. Will Hazell, "UK has 'no plan B' if Trump pulls out of NATO," *Telegraph*, 24 February 2024, available at: https://www.telegraph.co.uk/news/2024/02/24/nato-trump-defence-us-republicans-attack/ (last accessed 16/05/24).
10. Lisa Haseldine, "How can Germany deploy a tank battalion without any tanks?" *Spectator*, 27 January 2024, available at: https://www.spectator.co.uk/article/the-dire-state-of-germanys-army/ (last accessed 16/05/24).
11. A personal conversation in December 2023 between one of the authors and a leading figure close to Germany's Chancellor.
12. This remark was reported in Henry Samuel, "Sarkozy accused of being 'Russian influencer' for remarks on Ukraine War," *Telegraph*, 17 August 2023, available available at: https://www.telegraph.co.uk/world-news/2023/08/17/nicolas-sarkozy-accused-russian-influencer-ukraine-war/ (last accessed 16/05/24).
13. A personal conversation between one of the authors and a senior French figure close to the Elysee.
14. The *Zeitenwende* speech took place on 27 February 2022, five days after the Russian invasion of Ukraine. The main points of the speech were: Berlin would spend an additional €100 billion on defence and Germany would in future spend 2% GDP on defence. Since 2022 those commitments have been softened.
15. The dangers of such virtue signalling are not just confined to government. In July 2023, Andrew Griffith, the City minister, and James Cartlidge, the defence minister, warned that Britain's long-term security was being put at risk by City investors shunning investment in defence companies due to ethical concerns. See Hannah Boland, "Britain's security at risk from virtue-signalling banks, ministers warn," *Telegraph*, 30 July 2023, available at: https://www.telegraph.co.uk/business/2023/07/30/banks-risk-britain-security-avoid-defence-industry/ (last accessed 16/05/24).
16. BBC Daily Politics, "Autumn Statement Special," BBC TV, 5 December 2013.

2. A FORCE FOR GOOD?

1. Blair's Chief of Staff Jonathan Powell wrote about the Chicago Speech, "Tony was firmly in the internationalist and the idealist camp ... His argument was that we could no longer ignore what happens in other countries but had an interest and a duty to intervene if people were being suppressed by their rulers." Jonathan Powell, *The New Machiavelli: How to Wield Power in the Modern World*, London: Bodley Head, 2010, p. 263.
2. In response to the book's questionnaire.
3. "Number of fatalities among Western coalition soldiers involved in the execution of Operation Enduring Freedom from 2001 to 2021," Statista.com, available at: https://www.statista.com/statistics/262894/western-coalition-soldiers-killed-in-afghanistan/ (last accessed 16/05/24).
4. "Number of civilian deaths in Afghanistan from 2007 to 2020, by responsible party," Statista.com, available at: https://www.statista.com/statistics/269037/responsible-for-civilian-deaths-in-afghanistan/ (last accessed 16/05/24).
5. "Remarks by President Biden on Afghanistan," The White House, 16 August 2021, available at: https://www.whitehouse.gov/briefing-room/speeches-remarks/2021/08/16/remarks-by-president-biden-on-afghanistan/ (last accessed 16/05/24).
6. BBC Reality Check Team, "Biden on Afghanistan Fact-Checked," BBC, 19 August 2021, available at: https://www.bbc.com/news/58243158 (last accessed 16/05/24).
7. The 5 December 2001 Bonn Agreement was made between various Afghan anti-Taliban political factions and overseen by the international community engaged in Afghanistan. It established a road map and timetable for establishing peace and security, the reconstruction of Afghanistan, institution building and human rights.
8. "Remarks by President Biden on the Drawdown of U.S. Forces in Afghanistan," The White House, 8 July 2021, available at: https://www.whitehouse.gov/briefing-room/speeches-remarks/2021/07/08/remarks-by-president-biden-on-the-drawdown-of-u-s-forces-in-afghanistan/ (last accessed 11/05/24).
9. Martin Ewans, *Securing the Indian Frontier in Central Asia: Confrontation and Negotiation, 1865–1895*, London: Routledge, 2012, p. 70.

10. "Hostilities in the Gaza Strip and Israel: Flash Update 116," 12 February 2024, available at: https://www.unocha.org/publications/report/occupied-palestinian-territory/hostilities-gaza-strip-and-israel-flash-update-116 (last accessed 16/05/24).

3. SMOKE AND ERRORS

1. In response to the book's questionnaire.
2. In response to the book's questionnaire.
3. Peter Walker, "Michael Gove apologises for mistakes by government during Covid crisis," *Guardian*, 28 November 2023, available at: https://www.theguardian.com/uk-news/2023/nov/28/michael-gove-apologises-for-mistakes-by-government-during-covid-crisis (last accessed 16/05/24).
4. Lionel Shriver, "The government could tackle immigration—if it really wanted to," *Spectator*, 2 December 2023, available at: https://www.spectator.co.uk/article/the-government-could-tackle-immigration-if-it-really-wanted-to/ (last accessed 16/05/24).
5. Nick Timothy, "Britain must take back control and kick its addiction to immigration," *Telegraph*, 21 May 2023, available at: https://www.telegraph.co.uk/news/2023/05/21/britain-must-take-back-control-kick-addiction-immigration/ (last accessed 16/05/24).
6. Lawrence Freedman, *Strategy: A History*, Oxford: Oxford University Press, 1998.
7. "A Strong Britain in an Age of Uncertainty: The National Security Strategy," London: HMSO, 2010, p. 5.
8. "The National Security Strategy of the United Kingdom: Security in an interdependent world," London: HMSO, 2008, p. 5.
9. "Integrated Review Refresh 2023: Responding to a more contested and volatile world,", HM Government 2023, p. 11. Available at: https://assets.publishing.service.gov.uk/media/641d72f45155a2000c6ad5d5/11857435_NS_IR_Refresh_2023_Supply_AllPages_Revision_7_WEB_PDF.pdf (last accessed 12/05/24).
10. Ibid., p. 8.
11. P. Cornish, J. Lindley-French and C. Yorke, *Strategic Communications and National Strategy*, London: Chatham House, 2011.

12. "The UK Government Resilience Framework," HM Government, 4 December 2023. Available at: https://www.gov.uk/government/publications/the-uk-government-resilience-framework/the-uk-government-resilience-framework-html (last accessed 12/05/24).

4. ENDS, WAYS, AND HAS-BEENS?

1. Jonathan Beale and Jacqueline Howard, "What we know about strikes on Houthis and strategy behind them," BBC News, available at: https://www.bbc.com/news/world-middle-east-67955727 (last accessed 16/05/24).
2. See Malcolm Chalmers, "A Welcome Refreshment? Implications of the Spring Budget for UK Defence," RUSI Policy Brief, 17 May 2023, available at: https://www.rusi.org/explore-our-research/publications/policy-briefs/welcome-refreshment-implications-spring-budget-uk-defence (last accessed 16/05/24).
3. Matthew Harries, "Is the UK Capable of Maintaining its Nuclear Arsenal?" *Prospect*, 16 April 2022, available at: https://www.prospectmagazine.co.uk/politics/38577/is-the-uk-capable-of-maintaining-its-nuclear-arsenal (last accessed 12/05/24).
4. Dan Sabbagh, "Top issues in Grant Shapps' in-tray as new Defence Secretary," *Guardian*, 31 August 2023, available at: https://www.theguardian.com/politics/2023/aug/31/the-top-issues-in-grant-shapps-in-tray-as-new-uk-defence-secretary (last accessed 12/05/24).
5. In response to the book's questionnaire.
6. "The Equipment Plan 2023–2033," National Audit Office, 2023, available at: https://www.nao.org.uk/wp-content/uploads/2023/12/The-Equipment-Plan-20232033-Summary.pdf (last accessed 12/05/24).
7. For an excellent explanation of Britain's role in NATO defence see Ed Arnold, "The UK Contribution to Security in Northern Europe," RUSI Policy Brief, London: RUSI, 2023.
8. Nectar Gan, "Xi Jinping vows to make China's military a 'great wall of steel' in first speech of new presidential term," CNN, 13 March 2023, available at: https://edition.cnn.com/2023/03/13/china/china-xi-jinpong-first-speech-third-term-intl-hnk/index.html (last accessed 12/05/24).
9. Dan Sabbagh, "UK vulnerable to potential missile or drone attack, says

military chief," *Guardian*, 14 September 2023, available at: https://www.theguardian.com/uk-news/2023/sep/14/uk-vulnerable-to-potential-missile-or-drone-attack-says-military-chief (last accessed 12/05/24).

10. Douglas Barrie and Robert Wall, "UK Defence Command Paper aims to provide marching orders for industry," Military Balance Blog, IISS, 28 July 2023, available at: https://www.iiss.org/online-analysis/military-balance/2023/07/uk-defence-command-paper-aims-to-provide-marching-orders-for-industry/ (last accessed 12/05/24).
11. Rick Haythornthwaite, "Agency and Agility: Incentivising people in a new era—a review of UK Armed Forces incentivisation," Ministry of Defence, 19 June 2023, available at: https://www.gov.uk/government/publications/agency-and-agility-incentivising-people-in-a-new-era-a-review-of-uk-armed-forces-incentivisation (last accessed 12/05/24).
12. See "Report: The Future War, Strategy and Technology Conference," October 2023, available at: https://thealphengroup.com/2023/11/26/the-future-war-strategy-and-technology-conference/ (last accessed 12/05/24).
13. As NATO states, "The 1967 'Report of the Council on the Future Tasks of the Alliance,' also known as the Harmel Report, was a seminal document in NATO's history. It reasserted NATO's basic principles and effectively introduced the notion of deterrence and détente, setting the scene for NATO's first steps toward a more cooperative approach to security issues that would emerge in 1991." "Harmel Report," 1 July 2022, NATO, available at: https://www.nato.int/cps/en/natohq/topics_67927.htm (last accessed 12/05/24).
14. "Report: The Future War and Deterrence Conference," October 2022, available at: https://thealphengroup.com/2022/10/31/the-future-war-and-deterrence-conference-report/ (last accessed 15/05/24).
15. Available at parliament.uk: https://api.parliament.uk/historic-hansard/commons/1936/feb/24/armaments-industry-profits#S5CV0309P0_19360224_HOC_175 (last accessed 12/05/24).

5. A RETURN TO STRATEGY

1. "Air Force Doctrine Publication 1," US Air Force, 10 March 2021, available at: https://www.doctrine.af.mil/Portals/61/documents/AFDP_1/AFDP-1.pdf (last accessed 16/05/24).

2. NATO Warsaw Summit Communiqué, 9 July 2016, available at: https://www.nato.int/cps/en/natohq/official_texts_133169.htm (last accessed 16/05/24).
3. James Holland, *Normandy 44*, London: Penguin, 2020.
4. *The Oxford English Dictionary*, Oxford: Oxford University Press, 1989, p. 285.
5. "The UK Government Resilience Framework: 2023 Implementation Update," Cabinet Office, 4 December 2023, available at: https://www.gov.uk/government/publications/the-resilience-framework-2023-implementation-update/the-uk-government-resilience-framework-2023-implementation-update-html (last accessed 16/05/24).

6. THE UTILITY OF (BRITISH) FORCE

1. "Defence Command Paper 2023: Defence's response to a more contested and volatile world," Ministry of Defence, 18 July 2023, p. 13, available at: https://www.gov.uk/government/publications/defence-command-paper-2023-defences-response-to-a-more-contested-and-volatile-world (last accessed 17/05/24).
2. See "NATO military leadership addresses new era of collective defence," NATO, 19 January 2023, available at: https://www.nato.int/cps/en/natohq/news_210707.htm (last accessed 17/05/24).
3. Trevor Taylor and Andrew Curtis, "Management of Defence After the Levine Report: What Comes Next?" RUSI, 9 September 2020, https://www.rusi.org/explore-our-research/publications/occasional-papers/management-defence-after-levene-reforms-what-comes-next (last accessed 17/05/24).
4. The term "Enshitification" was coined in 2022 by Cory Doctorow.
5. Nick Gutteridge, "Lord Cameron says the world is the most 'dangerous' it has been in decades," Yahoo News, January 14, 2024, available at: https://uk.news.yahoo.com/lord-cameron-says-world-most-130806476.html
6. Figures from the Taxpayer's Alliance, available at: https://www.taxpayersalliance.com/new_bumper_book_of_government_waste_exposes_120_billion_of_wasteful_spending (last accessed 17/05/24).
7. See Wyn Bowen and G. Chapman, "The UK, Nuclear Deterrence and

a Changing World," Freeman Air and Space Institute, King's College London, December 2022, p. 13, available at: https://www.kcl.ac.uk/warstudies/assets/kcl-fasi-paper13-uk-nuclear-deterrence-changing-world-web.pdf (last accessed 17/05/24).

8. "MoD Departmental Resources: 2023," Ministry of Defence, 30 November 2023, Section 3, Figure 1, available at: https://www.gov.uk/government/statistics/defence-departmental-resources-2023/mod-departmental-resources-2023 (last accessed 17/05/24).
9. "Evidence Summary: The Drivers of Defence Cost Inflation," Ministry of Defence, February 2022, available at: https://www.gov.uk/government/publications/evidence-summary-the-drivers-of-defence-cost-inflation (last accessed 17/05/24).
10. Dan van der Vat, "Field Marshal Lord Carver: Obituary," *Guardian*, 12 December 2001, available at: https://www.theguardian.com/news/2001/dec/12/guardianobituaries (last accessed 17/05/24).
11. Matthew Harries, "Is the UK capable of maintaining its nuclear deterrent?" *Prospect*, 16 April 2022, available at: https://www.prospectmagazine.co.uk/politics/38577/is-the-uk-capable-of-maintaining-its-nuclear-arsenal (last accessed 17/05/24).
12. See "'Out of Control': Cost of Britain's nukes rose by 62% in 2023," Campaign for Nuclear Disarmament, 8 December 2023, available at: https://cnduk.org/out-of-control-cost-of-britains-nukes-rose-by-62-in-2023/ (last accessed 17/05/24).
13. Claire Mills and Esme Kirk-Wade, "The cost of the UK's strategic nuclear deterrent", Research Briefing, House of Commons Library, 3 May 2023, available at: https://researchbriefings.files.parliament.uk/documents/CBP-8166/CBP-8166.pdf (last accessed 17/05/24).

7. BELGIUM WITH NUKES?

1. Steven Edgington, "Army at risk of becoming 'static land force', chief tells generals," *Daily Telegraph*, 24 February 2024, available at: https://www.telegraph.co.uk/news/2024/02/24/army-risk-becoming-static-land-force-chief-tells-generals/ (last accessed 27/05/24).
2. In correspondence with Julian Lindley-French.
3. In correspondence with Julian Lindley-French.
4. "British Armed Forces—Statistics and Facts," Statista.com, 1 March

2024, available at: https://www.statista.com/topics/4219/armed-forces-of-the-united-kingdom/#topicOverview (last accessed 18/05/24).

8. TOMMY

1. Susan Ratcliffe (ed.), *Oxford Essential Quotations*, Oxford: Oxford University Press, 2016, available at: https://www.oxfordreference.com/display/10.1093/acref/9780191826719.001.0001/q-oro-ed4-00000981 (last accessed 18/05/24).
2. Cited in Ben Riley-Smith, "MoD to be denied funding boost in Budget," *Telegraph*, 26 February 2024, available at: https://www.telegraph.co.uk/politics/2024/02/26/mod-to-be-denied-funding-boost-in-budget/ (last accessed 18/05/24).
3. In response to the book's questionnaire.
4. Jonathan Powell, *The New Machiavelli: How to Wield Power in the Modern World*, London: The Bodley Head, 2010, p. 55.
5. Sir Hew Strachan, "The Lost Meaning of Strategy," *Survival*, Vol. 47, No. 3, Autumn 2005, p. 52.
6. Sir Lawrence Freedman, *Strategy: A History*, Oxford: Oxford University Press, 2013, p. xi.
7. Winston Churchill, *The Second World War: The Gathering Storm*, Vol. 6, London: Houghton, 1948.
8. Rudyard Kipling, "Tommy," available at: https://www.poetry.com/poem/33632/tommy (last accessed 18/05/24).
9. See "PM announces 'turning point' in European security as UK set to increase defence spending to 2.5% by 2030," Press Release, 10 Downing Street, 23 April 2024, available at: https://www.gov.uk/government/news/pm-announces-turning-point-in-european-security-as-uk-set-to-increase-defence-spending-to-25-by-2030#:~:text=%E2%80%9CToday%20is%20a%20turning%20point,at%20home%20and%20stronger%20abroad.%E2%80%9D (last accessed 18/05/24).
10. Niccolo Machiavelli, *The Prince*, Oviedo: Entrecacias, 2021.

BIBLIOGRAPHY

Source Documents (All Crown Copyright)

"Defence Command Paper 2023: Defence's response to a more contested and volatile world," Ministry of Defence, 18 July 2023.

"Global Britain in a Competitive Age: the Integrated Review of Security, Defence, Development and Foreign Policy," HM Government, 2021.

"Responding to a more contested and volatile world: the Integrated Review Refresh 2023," HM Government, 2023.

The National Risk Register 2022.

Sources

Bobbit, Philip, (2003), *The Shield of Achilles: War, Peace and the Course of History*, London: Penguin.

Bowen, Wyn & Chapman, G. (2022), *The UK, Deterrence and a Changing World*, London: King's College London.

Bracken, Paul (2013), *The Second Nuclear Age: Strategy, Danger and The New Power Politics*, New York: St Martin's Press.

Carr, E.H. (1964 Ed.), *The Twenty Years' Crisis, 1919-1939*, New York: Harper & Row.

Churchill, Winston S. (1948), *The Second World War: The Gathering Storm*, Vol. 6, London: Houghton Mifflin.

Cornish, Paul (1996), *British Military Planning for the Defence of Germany 1945–50*, London: MacMillan.

Diamond, Jared (2019), *Upheaval: Turning Point for Nations*, London: Little Brown.

BIBLIOGRAPHY

Dimbleby, Jonathan (2015), *The Battle of the Atlantic: How the Allies Won the War*, London: Penguin.

Dobbins, James; Jones, Seth G.; Runkle, Benjamin; Mohandas, Siddarth (2009), *Occupying Iraq: A History of the Coalition Provisional Authority*, Santa Monica: RAND.

Freedman, Lawrence (1989), *The Evolution of Nuclear Strategy*, Second Edition, London: MacMillan.

——— (1998), "The Revolution in Strategic Affairs," IISS Adelphi Paper 318, Oxford: Oxford University Press.

——— (2013), *Strategy: A History*, Oxford: Oxford University Press.

——— (2023), *Command: The Politics of Military Operations from Korea to Ukraine*, London: Allen Lane.

French, David (2012), *Army, Empire and Cold War: The British Army and Military Policy 1945-1971*, Oxford: Oxford University Press.

Fuller J.F.C. (2011 ed.), *Tanks in the Great War 1914–1918*, Uckfield: Naval and Military Press.

Galbraith J.K. (1983), *The Anatomy of Power*, Boston: Houghton Mifflin.

Goodhart, David (2017), *The Road to Somewhere: The Populist Revolt and the Future of Politics*, London: Hurst.

Hackett, John (1982), *The Third World War: The Untold Story*, New York: MacMillan.

Halliday, Fred (1989 Ed.), *The Making of the Second Cold War*, London: Verso.

Handel, Michael I., Ed., (2001, Third Edition), *Masters of War: Classical Strategic Thought*, London: Routledge.

Hart, Peter (2008), *1918: A Very British Victory*, London: Phoenix.

Herman, Arthur (2004), *To Rule the Waves: How the British Navy Shaped the Modern World*, London: Hodder & Stoughton.

Holland, James (2020), *Normandy '44*, London: Penguin.

Howard, Michael (2000), *The Invention of Peace*, Profile: London.

Huntingdon, Samuel J. (1996), *The Clash of Civilisations and the Remaking of World Order*, New York: Simon and Schuster.

Hurd, Douglas (2010), *Choose Your Weapons: The British Foreign Secretary 200 Years of Argument, Success and Failure*, London: Weidenfeld and Nicholson.

Husain, Amir (2017), *The Sentient Machine: The Coming Age of Artificial Intelligence*, New York: Scribner.

Husain, Amir, Allen, John R. (2019), *Hyperwar: Conflict and Competition in the AI Century*, Austin: SparkCognition Press.

Jenkins, Roy (2001), *Churchill*, London: Pan.

Johnson, Andrew (Ed.) (2014), *Wars in Peace: British Military Operations since 1991*, London: RUSI.

Kampfner, John, (2003), *Blair's Wars*, London: Simon and Schuster.

Keegan, John (2005), *The Iraq War*, Ottawa: Vantage Canada.

Kennedy, Paul (1993), *Preparing for the Twenty-First Century*, London: Harper Collins.

——— (1987), *The Rise and Fall of the Great Powers: Economic Change and Military Conflict from 1500 to 2000*, New York: Random House.

——— (2004 ed.), *The Rise and Fall of British Naval Mastery*, London: Penguin.

Kerbaj, Richards (2022), *The Secret History of Five Eyes*, London: Blink.

Kilcullen, David (2019), *The Accidental Guerrilla: Fighting Small Wars in the Midst of a Big One*, New York: Oxford University Press.

Kissinger, Henry (2001), (1994), *Diplomacy*, New York: Touchstone.

——— *Does America Need a Foreign Policy? Towards a Diplomacy for the Twenty-First Century*, New York: Touchstone.

——— (2014), *World Order: Reflections on the Character of Nations and the Course of History*, London: Allen Lane.

LaFeber, Walter (1991), *America, Russia and the Cold War 1945-1990*, 6th Edition, New York: McGraw-Hill.

Ledwidge, F. (2011), *Losing Small Wars: British Military Failure in Iraq and Afghanistan*, New Haven, CT: Yale University Press.

Liddell Hart, Basil (1921), "The 'Man-in-the-Dark' Theory of War: The Essential Principles of Fighting Simplified and Crystallized into a Definite Formula," *National Review* 75.

——— (1991 ed.), *Strategy*, London: Meridian.

Lindley-French J. (2023), *NATO: The Enduring Alliance*, London: Routledge.

——— (2016), *Demons and Dragons: The New Geopolitics of Terror*, London: Routledge.

Lindley-French J.; Allen, J.; Hodges, F. (2022), *Future War and the Defence of Europe*, Oxford: Oxford University Press.

Lindley-French, J. and Boyer, Y. (2012), *The Oxford Handbook of War*, Oxford: Oxford University Press.

Lukes, Steven (2005 ed.), *Power: A Radical View*, London: Palgrave Macmillan.

Massie, Robert K. (2007), *Castles of Steel: Britain, Germany and the Winning of the Great War at Sea*, London: Vintage.

McCarthy, Helen (2011), *The British People and the League of Nations*, Manchester: Manchester University Press.

Meyer, C. (2005), *DC Confidential*, London: Weidenfeld & Nicolson.

Picketty, Thomas (2014), *Capital in the Twenty-First Century*, Harvard: Harvard University Press.

Powell, Jonathan (2010), *The New Machiavelli: How to Wield Power in the Modern World*, London: Bodley Head.

Richards, David (2014), *Taking Command*, London: Headline.

Shirreff, Richard (2016), *2017: War with Russia*, London: Hodder and Stoughton.

Smith, General Sir Rupert (2005), *The Utility of Force: The Art of War in the Modern World*, London: Allen Lane.

INDEX

INDEX

INDEX

INDEX

INDEX

INDEX